THE ADHESION MOLECULE *FactsBook*

Other books in the FactsBook Series:

A. Neil Barclay, Albertus D. Beyers, Marian L. Birkeland, Marion H. Brown, Simon J. Davis, Chamorro Somoza, Alan F. Williams
The Leucocyte Antigen FactsBook

Robin E. Callard and Andy J. H. Gearing
The Cytokine FactsBook

Steve Watson and Steve Arkinstall
The G-Protein Linked Receptor FactsBook

Ed Conley
The Ion Channel FactsBook

Shirley Ayad, Ray Boot-Handford, Martin J. Humphries, Karl E. Kadler and Adrian Shuttleworth
The Extracellular Matrix FactsBook

THE ADHESION MOLECULE *FactsBook*

Rod Pigott
*British Bio-technology,
Oxford, UK*

Christine Power
*Glaxo Institute for Molecular Biology,
Geneva, Switzerland*

Academic Press
Harcourt Brace & Company, Publishers
LONDON SAN DIEGO NEW YORK BOSTON
SYDNEY TOKYO TORONTO

This book is printed on acid-free paper

ACADEMIC PRESS LIMITED
24–28 Oval Road
LONDON NW1 7DX

United States Edition published by
ACADEMIC PRESS INC.
San Diego, CA 92101

Copyright © 1993 by
ACADEMIC PRESS LIMITED

All rights reserved
No part of this book may be reproduced in any form by photostat, microfilm, or by any other means, without written permission from the publishers

A catalogue record for this book is available from the British Library

ISBN 0–12–555180–0

Designed by Eric Drewery and Adrian Singer
Typeset by Columns Design and Production Services Ltd, Reading
Printed and bound in Great Britain by
Mackays of Chatham PLC, Chatham, Kent

Contents

Preface VII
Abbreviations VIII

Section I   THE INTRODUCTORY CHAPTERS

Chapter 1
Introduction 2

Chapter 2
The Cadherin Family 6

Chapter 3
The Integrin Family 9

Chapter 4
The Immunoglobulin Superfamily 13

Chapter 5
The Selectin Family 17

Section II   THE ADHESION MOLECULES

Adhesion molecules

AMOG	22	Leukocyte adhesion receptor	
B-Cadherin	23	p150, 95	89
CD2	26	LFA-1	93
CD4	28	LFA-3	98
CD8	31	L-Selectin	100
CD22	34	M-Cadherin	103
CD23	36	Mac1	105
CD35	39	Myelin-associated	
CD36	43	glycoprotein	109
CD44	46	N-Cadherin	112
CEA	49	NCAM	115
CELL-CAM 105	52	Neuroglian	120
Contactin	54	Ng-CAM	123
E-Cadherin	57	P-Cadherin	126
E-Selectin	60	PECAM-1	129
F3	64	P-Selectin	132
Fasciclin I	67	Platelet glycoprotein	
Fasciclin II	69	GP IIb–IIIa	135
Fasciclin III	72	Platelet glycoprotein	
ICAM-1	74	GPIb–IX complex	140
ICAM-2	77	R-Cadherin	145
Integrin α6β4	79	T-Cadherin	148
Integrin α7β1	83	TAG-1	151
L1	86	VCAM-1	154

Contents

Vitronectin receptor _____ 158
VLA-1 _____ 165
VLA-2 _____ 170
VLA-3 _____ 173
VLA-4 _____ 176
VLA-5 _____ 181
VLA-6 _____ 184

Index _____ 187

Preface

The authors wish to thank all those who reviewed the entries for this volume for their comments and suggestions. In compiling the entries we have become acutely aware of the pace at which new information is becoming available. It is therefore the intention that this volume should evolve through the publication of updated editions. We would encourage comments from readers on any errors, omissions and improvements which may subsequently be incorporated. These can be sent to: the Editor, Adhesion Molecule FactsBook, Academic Press, 24–28 Oval Road, London, NW1 7DX, UK.

Left: *Rod Pigott*, Right: *Christine Power*

Abbreviations

CD	Cluster of differentiation
CLA	Cutaneous lymphocyte antigen
CRP	Complement regulatory protein
CTL	Cytotoxic T lymphocyte
Cy	Cytoplasmic domain
EBNA2	Epstein–Barr virus nuclear antigen
EBV	Epstein–Barr virus
ECF-A	Eosinophil chemotactic factor A
ECM	Extracellular matrix
EGF	Epidermal growth factor
Fg	Fibrinogen
fMLP	f-met-leu-phe
FN	Fibronectin
GM-CSF	Granulocyte monocyte colony stimulating factor
GPI	Glycosyl phosphatidylinositol
HEV	High endothelial venule
HUVEC	Human umbilical vein endothelial cells
IFN	Interferon
Ig	Immunoglobulin
IL	Interleukin
IMF	Integrin modulating factor
IP3	Inositol triphosphate
LAD	Leukocyte adhesion deficiency
Lec	Lectin
LHR	Lymphocyte homing receptor
LPS	Lipopolysaccharide
NK	Natural killer
PAF	Platelet activating factor
PCR	Polymerase chain reaction
PGE	Prostaglandin E
PMA	Phorbol myristate acetate
PMN	Polymorphonuclear leukocyte
PPME	Polyphosphomannan ester
SCR	Short consensus repeat
SDS PAGE	Sodium dodecyl sulphate polyacrylamide gel electrophoresis
TCR	T cell receptor
TGF	Transforming growth factor
TM	Transmembrane
TNF	Tumour necrosis factor
vWF	von Willebrand factor

Section I

THE INTRODUCTORY CHAPTERS

1 Introduction

SCOPE OF THIS BOOK

Our aim in producing this book was to provide concise information on the structure, function and biology of cell adhesion molecules in a common and easily accessible format. Such a format necessarily dictates that the information included for each entry is restricted. It has however been our endeavour to include the necessary references such that the reader can readily access any further information they may require.

SELECTION OF ENTRIES

The criteria for selection of the entries were that the primary sequence of the molecule was available and that there was good evidence that one of the functions of the molecule was to increase the affinity of a cell, either for another cell, or for components of the extracellular matrix. At its most conclusive this evidence consists of a demonstration that when a putative adhesion molecule is artificially expressed in a cell it confers upon the cell a novel adhesive phenotype and that the new adhesive interaction can be inhibited by specific antibodies. Rigorous data of this nature has been obtained for a minority of the entries in this book. The majority of molecules have been assigned the title cell adhesion molecule taking into consideration a combination of pieces of evidence, pre-eminent amongst them the peturbation of cell–cell interactions by specific antibodies and the demonstration of a structural relationship to proven cell adhesion molecules.

The selection of entries was further complicated by the fact that few adhesion molecules function purely to increase the affinity of one cell for another. It is becoming increasingly clear that a large number of molecules also have a role in cell signalling. Dissecting these functions such that the relative importance of a molecule as a cell signalling or cell adhesion molecule can be determined may be extremely difficult.

Finally in a rapidly moving field it is inevitable that we will have omitted some molecules. Putative adhesion molecules, for which we felt there was insubstantial evidence confirming their role at the time of writing, will no doubt have been characterized further to warrant their inclusion.

ORGANIZATION OF THE DATA

Name

The entries are identified by the most commonly used name which in some cases will be a CD (cluster of differentiation) classification. Alternative names are given and where appropriate, the names of any species homologues.

Family

The majority of the adhesion molecules which have been identified fall into one of four families; the immunoglobulin superfamily (Ig-SF), the integrins, the selectins and the cadherins. The integrin entries appearing in Section II are listed in alphabetical order according to the most commonly used name, and includes all heterodimers known to date. Since most β subunits are known to associate with more than one α subunit, details for the promiscuous β subunit will only be given in one entry, usually the most widely known association, to which readers will be refered. A number of orphan molecules are categorized according to any structural motifs they share with other molecules.

Cellular distribution
Cell-types which have been clearly demonstrated to express the molecule are listed. Where this is particularly extensive it may be incomplete. A more detailed distribution of those molecules with a CD classification can be found in *Leucocyte Typing IV* (Knapp, W. *et al.* 1989, Oxford University Press, Oxford, UK). In many cases a complete survey of all tissues will not have been carried out. It should not therefore be assumed that because a given cell-type is not mentioned it does not express the molecule.

Function
The major functional roles of the molecule are given including any involvement in signal transduction.

Ligands
Includes the principal ligands (if known) and any key structural features required for binding.

Gene structure
Where the organization of the genomic DNA is known this is represented schematically. Introns and exons are not drawn to scale. Exon numbers are related to the protein structural domains where appropriate. Alternatively spliced exons are cross-hatched, untranslated sequences are stippled. See Table 1 for a key.

Gene location and size
The chromosomal location of the gene is given together with the approximate size.

Structure
The structure of the molecule is presented schematically. The domains within the molecule are identified as shown in Table 1.

Alternative forms
Includes any forms produced by alternative splicing, genetic polymorphisms or as a result of any cell-type specific post-translational modifications.

Molecular weights
The relative molecular weight (M_r) of the polypeptide has been calculated from the published sequence minus the signal and any propeptide sequences. For proteins with a GPI-anchor the M_r given is that of the mature polypeptide where the potential or actual site of attachment is known, otherwise it is of the unprocessed polypeptide. The value given for SDS–PAGE is that observed under reducing conditions. As there is often considerable variation within the literature a range of values may be given. Where there is no published M_r, the figure given is based on the weight of the polypeptide core plus 3000 M_r per predicted N-linked oligosaccharide residue.

Introduction

Table 1. *Keys to the schematic diagrams.*

	GENE STRUCTURE		PROTEIN STRUCTURE
☐	Exon	■	Transmembrane domain
▨	Non coding sequence	▦	EGF-like domain
▧	Alternatively spliced exon	▨	Fibronectin type III
S	Signal sequence	▤	Cytoplasmic domain
IG	Immunoglobulin domain	▨	Leucine-rich repeat
TM	Transmembrane domain	▨	Cysteine-rich domain
CY	Cytoplasmic domain	▨	Divalent cation binding site
TI	Translation initiation	▨	Variably expressed domain
LEC	Lectin domain	▤	'I' domain
EGF	EGF domain	▨	J or Hinge
CPP	Cleaved protein precursor	▨	Cartilage proteoglycan core and link
HIN	Hinge	▧	C2 Immunoglobulin-like domain
a	Alu repeat		
CYS	Cysteine rich domain	▨	V Immunoglobulin-like domain
CRP	Complement regulatory protein	▨	Complement regulatory-like domain
STOP	Stop codon	▨	Lectin-like domain

Sequences

The amino acid sequence is shown in the single letter amino acid code (Table 2). Numbering starts at the initiator methionine in the signal peptide, except in cases where only the mature protein sequence is known. The following structural features are identified if they occur;

End of signal sequence	:	↓
Cleavage sites:	:	▭
Divalent cation binding sites	:	shaded box
N-linked glycosylation sites	:	*
Transmembrane regions	:	underlined

In assigning domains we have aimed to achieve consistency within families. It is therefore possible that the boundaries will not coincide exactly with those originally assigned.

Where appropriate the sequences of any alternatively spliced variants are also given.

Database accession numbers

The PIR (Protein Identification Resource) and SWISSPROT accession numbers are given for protein sequences and the GENBANK/EMBL accession number for nucleic acid sequences. Where only a genomic sequence exists and it is spread across several entries more than one accession number is given.

References

It was not compatible with the format to provide a fully comprehensive reference list. However each entry includes a number of key references which are highlighted in bold. These are either recent reviews or recent papers which have been selected as a source of further relevant references.

Table 2. *Amino acid codes*

Amino acid	Single letter code	Three letter code
Alanine	A	Ala
Arginine	R	Arg
Asparagine	N	Asn
Aspartic acid	D	Asp
Cysteine	C	Cys
Glutamine	Q	Gln
Glutamic acid	E	Glu
Glycine	G	Gly
Histidine	H	His
Isoleucine	I	Ile
Leucine	L	Leu
Lysine	K	Lys
Methionine	M	Met
Phenylalanine	F	Phe
Proline	P	Pro
Serine	S	Ser
Threonine	T	Thr
Tryptophan	W	Trp
Tyrosine	Y	Tyr
Valine	V	Val

2 The Cadherin Family

The cadherins are a family of calcium-dependent adhesion molecules found both within and outside the nervous system. In humans at least 11 different cadherins have now been identified [1] which share a high degree of homology at the amino acid level (43–58%). Cadherin counterparts have also been cloned or purified from other species of mammal, birds and amphibians. *Drosophila* molecules with significant identities in amino acid sequence to vertebrate cadherins have also been detected [2].

FUNCTION

Cadherins generally mediate homotypic cell–cell adhesion although heterotypic binding between different cadherin molecules is possible. They thus act as both receptor and ligand. They are responsible for the selective cell–cell adhesion or cell sorting which is necessary to allocate different cell types to their proper positions during development. They also play a fundamental role in maintaining the integrity of multicellular structures. During embryonic morphogenesis, the expression of multiple members of the cadherin family is spatio-temporally regulated, and correlates with a variety of morphogenetic events that involve cell aggregation or disaggregation.

The role of cadherins in these processes has been demonstrated by the use of anti-cadherin monoclonal antibodies, which, when added to embryonic tissues, cause a severe distortion of structure which results in dissociation of the tissue into small clusters or even single cells. Similarly, Nagafuchi et al. [3] showed that transfection of E.cadherin cDNA into mouse L cells (which normally show little endogenous cadherin activity) resulted in the cells acquiring high Ca^{2+}-dependent aggregating activity. For detailed reviews the reader is referred to refs 4–7.

STRUCTURE

The cadherins are synthesized as a precursor polypeptide which requires a series of post-translational modifications (glycosylation, phosphorylation and proteolytic cleavage) to form a mature protein which is between 723 and 748 amino acids long. A schematic representation is shown in Fig. 1.

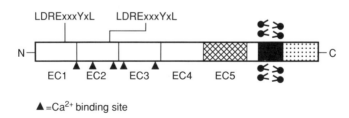

▲ = Ca^{2+} binding site

Figure 1. *Schematic representation of cadherin structure.*

The extracellular domain contains 3–5 internal repeats of approximately 110 amino acids (the number of repeats may vary depending on the criteria chosen for similarity). Repeats 1–3 each contain the putative Ca^{2+} binding site motif DXNDN or DXD, resembling the Ca^{2+} binding loops of the so-called EF-hand domains [8]. Another highly conserved cluster with the motif LDREXXXXYXL is also found within the first three repeats.

The N-terminal 113 amino acids which contain a conserved HAV sequence, have been shown to be important in ligand binding and specificity [9]. The extracellular domain is anchored to the cell membrane by a transmembrane domain of approximately 24 amino acids. The short cytoplasmic domain is the most highly conserved region of homology between cadherins and is particularly important for cadherin function.

REGULATION OF EXPRESSION AND SIGNAL TRANSDUCTION

Cadherins are differentially expressed during development and in adult organs. Since many cell types express multiple cadherin subclasses simultaneously (the combination differs with cell type), it can be inferred that the adhesion properties of individual cells are thus governed by varying the combinations of cadherins. There is also recent evidence to suggest that altered expression of cadherins may be involved in invasion and metastasis of tumour cells [10, 11]. However, at present little is known about the regulatory mechanisms controlling the differential expression of cadherins.

On the cell surface, cadherins tend to be concentrated at cell–cell junctions (zonula adherens junctions). Here they are structurally associated with cortical actin bundles (not stress-fibres) [12]. The cytoplasmic domain has also been shown to be associated with cytoplasmic proteins termed catenins α (102 kD), β (88 kD) and γ (80 kD) [13]. Deletion of the cytoplasmic domain destroys these interactions and also eliminates cadherin function [14]. Recently *src*, *yes* and *lyn* gene products of the *src* proto-oncogene family have been found expressed at zonula adherens junctions. [6] These kinases may be responsible for cadherin phosphorylation and this raises the possibility that cadherin mediated cell junctions might be used for intercellular signalling [15, 16].

The seven cadherins described in this book are those which fall into the E-, N- and P-subclasses [4] and for which complete sequence information is available. It is clear, however, that the list of cadherins identified to date is by no means complete, particularly for those expressed in the brain. Without doubt, polymerase chain reaction (PCR) techniques will prove invaluable in identifying new members of this family.

References
[1] Suzuki, S. et al. (1991) Cell Regul. 2, 261–270.
[2] Mahoney, P.A. et al. (1991) Cell 67, 853–868.
[3] Nagafuchi, A. et al. (1987) Nature 329, 341–343.
[4] Takeichi, M. (1988) Development 102, 639–655.
[5] Takeichi, M. (1990) Annu. Rev. Biochem. 59, 237–252.
[6] Takeichi, M. (1991) Science 251, 1451–1455.
[7] Takeichi, M. et al. (1990) Neurosci. Res. (suppl.) 13, S92–S96.
[8] Ringwald, M. et al. (1987) EMBO J. 6, 3647–3653.
[9] Nose, A. et al. (1990) Cell 61, 147–155.
[10] Behrens, J. et al. (1989) J. Cell Biol. 108, 2435–2447.
[11] Behrens, J. et al. (1992) Seminars Cell Biol. 3, 169–178.
[12] Hirano, S. et al. (1987) J. Cell Biol. 105, 2501–2510.

13 Ozawa, M. et al. (1989) EMBO J. 8, 1711–1717,
14 Ozawa, M. et al. (1990) Proc. Natl. Acad. Sci. USA 87, 4246–4250.
15 Matsuyoshi, N. et al. (1992) J. Cell Biol. 118, 703–714.
16 Saffell, J.L. et al. (1992) J. Cell Biol. 118, 663–670.
17 Kemler, R. (1992) Seminars Cell Biol. 3, 149–155.

3 The Integrin Family

The integrins are a family of heterodimeric membrane glycoproteins expressed on diverse cell types which function as the major receptors for extracellular matrix and as cell–cell adhesion molecules. As adhesion molecules they play an important role in numerous biological processes such as platelet aggregation, inflammation, immune function, wound healing, tumour metastasis and tissue migration during embryogenesis. In addition, there is now increasing evidence to implicate integrins in signalling pathways, transmitting signals both into and out from cells. For greater detail, the reader is referred to a number of excellent reviews 1–6.

STRUCTURE

All integrins consist of two non-covalently associated subunits, α and β. The integrins were originally classified into three subfamilies (β1 integrins or VLA proteins; β2 integrins or leucams and β3 integrins or cytoadhesins) in which a common β subunit was thought to associate with a number of different α subunits. However, this classification is now less rigid since to date at least 12 different α subunits and 8 β subunits have been identified. Furthermore, individual α subunits have been shown to associate with more than one type of β subunit. The known integrin subunit associations are shown in Fig. 1.

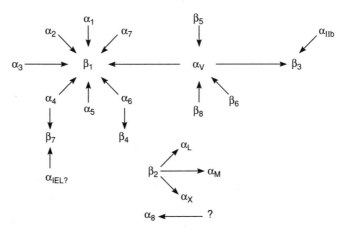

Figure 1. *Integrin subunit associations.*

Figure 2 shows a schematic representation of a typical integrin heterodimer. The β subunits (molecular weight range from 90 to 110 kD except β4 (210 kD)) all show strong homology at the amino acid level of between 40–48%. They all contain 56 conserved cysteines (except β4 which has 48) which are arranged in four repeating units. Biochemical analyses of the β3 subunit suggest that there is a large loop in the N-terminus stabilized by intrachain disulphide bonding with the first cysteine-rich repeat. The cytoplasmic domain is short (40–50 amino acids) the exception again being β4 which contains a long cytoplasmic domain of 1018 amino acids containing four fibronectin type III repeats. The functional significance of this is unknown at present.

The Integrin Family

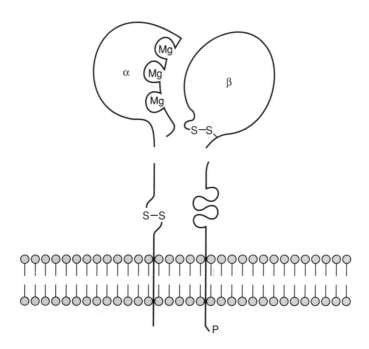

Figure 2. *Schematic representation of a typical integrin.*

The α subunits (molecular weight range from 150 to 200 kD) exhibit lower sequence homology than the β subunits. All contain seven repeating domains of 24-45 amino acids spaced 20–35 amino acids apart. The N-terminal of the α subunit contains 3 or 4 divalent cation binding sites. Subunits α1, α2, αL, αM and αX all contain an extra 200 amino acids inserted between repeats 2 and 3 known as the I-domain or inserted domain. The α subunits which do not contain the I-domain undergo cleavage into heavy and light fragments which are disulphide linked. However, the α4 subunit is an exception in that it does not contain an I-domain but undergoes cleavage to yield 70 and 80 kD fragments which are not disulphide linked.

LIGANDS

The integrins bind to a wide variety of extracellular matrix proteins (fibronectin, fibrinogen, laminin, collagen, thrombospondin, vitronectin and von Willebrand factor) and to members of the immunoglobulin superfamily such as ICAM-1, -2 and -3 and VCAM-1. Most members of the integrin family bind more than one ligand and exhibit different specificities depending on the cell type in which they are expressed. The existence of alternatively spliced forms differing in the extracellular domain may be important in altering ligand binding affinity.

The recognition site for some integrins is an RGD sequence on the extracellular matrix or platelet adhesion protein [7]. However, not all integrins bind this ligand sequence. Ligand binding is cation-dependent and requires both subunits. The cytoplasmic domains of both subunits also appear to be required for binding to cytoskeletal components, thus playing an important role in linking the extracellular matrix to cytoskeleton.

REGULATION OF INTEGRIN FUNCTION AND SIGNAL TRANSDUCTION

In general, adhesion mediated by integrins requires activation of the receptor by specific signals, for example thrombin (in the case of αIIb/β3) or T cell activation (α4β1). The activation of the receptor results in a conformational change which enables it to bind ligand. Ligand occupation may then trigger a series of intracellular events.

Recent work has shown that regulation of integrin function may involve phosphorylation of the β subunit cytoplasmic domain [8]. Indeed phosphorylation of the β2 integrins has been shown to induce homotypic adhesion [9]. Serine and tyrosine phosphorylation has been detected on the β4 subunit and β1, β3, β5 and β6 subunits all possess potential tyrosine phophorylation sites. The α subunit cytoplasmic domain does not appear to play a role in the regulation of ligand binding; however, it appears to be important in translating ligand binding into subsequent cellular events [6, 10].

It is interesting to note that multiple integrins which bind the same ligand are co-expressed on the same cell type. The apparent redundancy can be explained by the fact that differing cellular responses are mediated by different integrin cytoplasmic domains in response to common extracellular ligands. The existence of variant cytoplasmic domains on several integrin subunits, generated by alternative splicing, may therefore add even more versatility to intracellular signalling processes.

CLINICAL ASPECTS

The clinical significance of integrins has been highlighted by the discovery of a number of diseases in which there is an integrin defect. Leukocyte adhesion deficiency (LAD) is a rare inherited disorder in which key functions of leukocytes are impaired, notably the migration of neutrophils to sites of extravascular inflammation. Affected individuals suffer from recurrent and often fatal bacterial infections. It now appears that LAD results from a structural defect in the β2 subunit of the leukocyte integrins arising by abnormal splicing and a missense mutation [11]. Similarly, a defect or deficiency in platelet glycoprotein αIIb/β3 appears to be important in Glanzmann's thrombasthemia, a bleeding disorder characterized by a failure of platelet aggregation in response to agonist stimulation.

The involvement of specific integrins in various disease processes has been elucidated by the use of monoclonal antibodies. For example, monoclonal antibodies against β2 integrins have been shown to be effective in animal models of inflammation, septic shock and ischaemia-reperfusion injury [12]. Similarly, in vitro, RGD peptides have shown a number of diverse effects such as prevention of platelet aggregation [13], inhibition of tumour cell migration [14] and inhibition of osteoclast binding to bone [15]. Hence, as our knowledge of the roles of specific integrins and their respective ligands increases, this will lead to the development of numerous potentially useful therapeutic tools.

References

1. Albelda, S.M. and Buck, C.A. (1990) FASEB J. 4, 2868–2880.
2. Hynes, R.O. (1987) Cell 48, 549-554.
3. Larson, R.S. and Springer, T.A. (1990) Immunol. Rev. 114, 181–217.
4. Hemler, M.E. (1991) Annu. Rev. Immunol. 8, 365–400.
5. Ruoslahti, E. (1991) J. Clin. Invest. 87, 1–5.
6. Hynes, R.O. (1992) Cell 69, 11–25.
7. D'Souza, S. et al. (1991) Trends Biochem. Sci. 16, 246–250.
8. Hibbs, M.L. et al. (1991) Science 251, 1611–1613.
9. Chatila, T.A. et al. (1989) J. Cell Biol. 109, 3435–3444.
10. Chan, B.M.C. et al. (1992) Cell 68, 1051–1060.
11. Nelson, C. et al. (1992) J. Biol. Chem. 267, 3351–3357.
12. Rosen, H. and Gordon, S. (1989) Br. J. Exp. Pathol. 70, 385–394.
13. Cook, J.J. et al. (1989) Am. J. Physiol. 256, H1038–F1043.
14. Gehlsen, K.R. et al. (1988) J. Cell Biol. 106, 925–930.
15. Reinholt, F.P. et al. (1990) Proc. Natl Acad. Sci. USA 87, 4473–4475.

4 The Immunoglobulin Superfamily

Since the concept of the immunoglobulin superfamily (IgSF) was proposed in 1982[1], it has expanded to embrace over 70 members, including both single- and multi-gene representatives. The roles of the family members are multivarious, but are linked by the common theme of controlling cell behaviour. Such control is exerted by molecules acting as signal transducing receptors (for example the PDGF and IL1 receptors) or as intercellular adhesion molecules or, as is becoming increasingly apparent, as both. As would be expected of such molecules, with the exceptions of a single family member found intracellularly (the skeletal muscle C-protein) and three members which probably exist only as secreted proteins (the serum protein α1B-gp, the link protein of basement membranes and the proteoglycan, perlecan), they can all be found as cell surface molecules, although soluble isoforms also exist for a number of family members.

STRUCTURE

The complex array of structures found within the Ig superfamily is thought to have evolved from a single ancestral unit, the immunoglobulin fold, which has undergone divergence and duplication. This basic motif is found in species as diverse as insects and man. The immunoglobulin fold is composed of between 70 and 110 amino acid residues, which are organized into two parallel β sheets which, in the majority of cases, are stabilized by characteristic disulphide bonds (for excellent reviews see refs 2–4). In the immunoglobulin variable (V-type) domain, the β sandwich is formed by two β sheets, one of 3 and the other of 4 strands. Lying between these sheets are two additional domains, C' and C" which, in antibodies, form the second hypervariable loop. The constant region domain (C-type) is missing the two additional strands. The C-type domain has been subdivided into C1 and C2 (or H). Members of the C2 subtype are characterized by sequence similarities outside the regions of β strand and often have a reduced number of residues in the middle, D strand. Amongst adhesion molecules with C-type domains, it is the C2 subtype which is exclusively represented.

With the exception of a few highly conserved residues, there is an enormous variation in the primary structure of the members of the immunoglobulin superfamily, yet the tertiary structures are remarkably constant. It seems probable that this basic structure serves as a scaffold on which unique determinants can be displayed for recognition, either on the surfaces of the β sheets, or at the turns connecting the strands. The evolutionary pressure to maintain such a structure is thought to be that it represents a conformation which is protease resistant and is able to survive in the hostile extracellular environment.

The earliest recognition events mediated by the members of the IgSF, were probably between identical single domains on opposing cells (simple homophilic interactions). The increasing specialization and specificity of interactions of IgSF members has been generated through combinations of at least four different mechanisms: (1) by the diversification of individual domains to produce new molecules with different specificities; (2) by the duplication of domains to produce multidomain variants (to date the greatest number of Ig-like repeats has been found in the extracellular matrix glycoprotein perlecan, which has a total of 14); (3) by the combination of domains either with other Ig-like domains, or with different structural units such as the fibronectin type III repeat, as found in NCAM; (4) by association with other cell surface receptors. Members of the IgSF may themselves associate to form functional units, as in the case of CD8, or alternatively may be found as one half of a co-receptor pair as in the case of CD4 and the T cell receptor.

The production of multidomain molecules within the IgSF is made possible by the fact that in the overwhelming majority of cases, individual domains are coded for by individual exons. Each exon is constructed such that at the 3' end is a single base from the first codon for the domain immediately following and the 5' end has two bases from the last codon for the domain immediately preceding. Although it is not possible to determine the exact evolutionary sequence by which a multidomain molecule was generated from the single domain ancestral unit, clues can be found in the homologies present within a molecule. For example, the carcinoembryonic antigen (CEA) has a single V-domain, followed by six C2-domains. Domains 2, 4 and 6 are 70% identical, as are domains 3, 5 and 7. However, between these two sets the homology is only in the region of 25%. This suggests that there has been a double duplication of an ancestral gene encoding two domains [5]. In the vascular cell adhesion molecule (VCAM-1) domains 1 and 4 are clearly homologous, as are domains 2 and 5, and 3 and 6, suggesting that in this case, VCAM-1 was generated from duplication of an ancestral gene encoding three domains [6].

Notable exceptions to the one exon, one domain rule are all the immunoglobulin-like domains of NCAM and domain 1 of CD4, where each domain is encoded by two exons. This may reflect the possibility that the immunoglobulin fold was itself formed by the duplication of an ancestral domain [7]. Evidence to support such a theory has recently been provided by studies showing that a functionally important region of the fibroblast growth factor receptor forms the C-terminal half of an immunoglobulin-like domain and is encoded by a separate exon [8].

LIGANDS

The role of the immunoglobulin fold in cell adhesion is thought to have originated by the interaction of single, identical domains on opposing cells (homophilic adhesion) [4]. As multidomain molecules have evolved, this capacity for homophilic interactions has been maintained and such interactions are common amongst members of the immunoglobulin superfamily. In an extension to this mechanism interactions may also be mediated by non-identical family members, such as CD2 and LFA-3.

Within recent years other non-IgSF members have been identified as ligands. Amongst the most important of these are a number of members of the integrin family including LFA-1 (CD18/CD11a) and Mac-1 (CD18/CD11b) which bind ICAM-1 [9,10] and VLA-4 ($\alpha4/\beta1$) [11] and $\alpha4/\beta7$ [12] which bind VCAM-1.

A diverse range of additional counter receptors have been identified including a third ligand for ICAM-1, CD43 and components of the extracellular matrix including collagen and heparin, recognized by myelin associated glycoprotein and heparan sulphate recognized by NCAM.

FUNCTION

It has become increasingly obvious that molecules originally identified as adhesion molecules may have diverse functions and that in some instances the adhesion role may be of secondary importance. Accumulating evidence indicates that an exclusive role in cell adhesion may well be the exception rather than the rule, with the majority of molecules serving both as adhesion receptors and signal transducers. There exists therefore, a continuum, from molecules that function solely to increase the affinity of one cell for another, to those which are exclusively involved in cell signalling.

Multiplicity of function is well-illustrated by NCAM. Although NCAM was one of the earliest members of the IgSF to be assigned the title cell adhesion molecule, the evidence to support this claim is still not beyond dispute. It seems probable that in addition to any direct role in cell adhesion, NCAM is involved both in regulating the fundamental ability of cells to interact [14] and in signal transduction [15].

Immunoglobulin superfamily members are widely utilized in two areas involving complex interactions amongst a diverse array of cell types: during development and in the regulation of the immune system. Their role in development is particularly evident within the nervous system. A group of molecules with multiple immunoglobulin domains coupled to a varying number of repeats related to the fibronectin type III repeat, is involved in cell migration, the stimulation and inhibition of neurite outgrowth and the adhesion of neurites [16]. The fundamental importance of this group is reflected in the fact that homologues have been detected in species as diverse as insects and man. Cellular interactions within the immune system are no less complex than those found within the nervous system and here too IgSF members are widely deployed, including ICAMs 1 and 2, VCAM-1, CD4, CD8, LFA-3, CD2 and CD22 [17].

CONCLUSIONS

The complexity of the IgSF continues to increase due not only to the identification of new members, but also to the identification of new functions for existing members. Future studies in this field are likely to focus on the molecular analysis of the cell binding sites of IgSF members, on the roles and mechanisms of signal transduction and on interrelationships with other adhesion molecules.

References
1. Williams, A.F. (1982) J. Theoret. Biol. 98, 221–231.
2. Williams, A.F. et al. (1989) Cold Spring Harb. Symp. Mol. Biol. 54, 637–647.
3. Hunkapiller, T. and Hood, L. (1989) Adv. Immunol. 4, 1–63.
4. Williams, A.F (1987) Immunol. Today 8, 298–303.
5. Oikawa, S. et al. (1987) Biochem. Biophys. Res. Commun. 144, 634–642.
6. Polte, T. et al. (1991) DNA Cell Biol. 10, 349–357.
7. Bourgois, A. (1975) Immunochemistry 12, 873–876.
8. Yayon, A. et al. (1992) EMBO J. 11, 1885–1890.
9. Marlin, S.D and Springer, T. (1987) Cell 51, 813–819.
10. Diamond, M.S. et al. (1990) J. Cell Biol. 111, 3129–3139.
11. Elices, M.J. et al. (1990) Cell 60, 577–584.
12. Ruegg, C. et al. (1992) J. Cell Biol. 117, 179–189.

13 Rosenstein, Y. et al. (1991) Nature 354, 233–235.
14 Rutishauser, U. (1991) Seminars Neurosci. 3, 265–270.
15 Doherty, P. et al. (1991) Cell 67, 21–33.
16 Rathjen, F.G. and Jessell, T. (1991) Seminars Neurosci. 3, 297–307.
17 Springer, T.A. (1990) Nature 346, 425–434.

5 The Selectin Family

Despite the diminutive size of the family (three members) and the fact that protein sequences were only obtained in 1989, there is already an extensive literature on the selectins. The driving force behind this productivity is the strong evidence that E-, P- and L-selectin are involved in the inflammatory response and therefore represent novel therapeutic targets. Much of this literature has been the subject of a number of excellent recent reviews to which the reader is referred for greater detail [1,2].

STRUCTURE

The selectins are constructed from three types of protein domain. At the N-terminal is a domain related to the calcium-dependent (C-type) lectin domain, found in a range of proteins including serum glycoproteins and proteoglycans of the extracellular matrix [3]. This is followed by a domain related to a repeat first described in epidermal growth factor (EGF). The degree of homology between the three family members within these two domains is shown in Fig. 1. The EGF-domain is succeeded by a varying number of repeats of a domain found in complement regulatory proteins (CRP). As with the lectin-EGF-domains, the CRP repeats show a high degree of homology between family members. The transmembrane region is followed by a short cytoplasmic tail. The longest cytoplasmic domain (35 residues) is found in P-selectin, one function of which is to direct sorting to storage granules [4]. No function has yet been ascribed to the cytoplasmic regions of L- and E-selectin.

```
E-SELECTIN   WSYNTSTEAM TYDEASAYCQ QHYTHLVAIQ NKEEIHYLNS ILSYSHSYYW   50
P-SELECTIN   WTYHYSTKAY SWNISRKYCQ NRYTDLVAIQ NKNEIDYLNK VLPYYSSYYW   50
L-SELECTIN   WTYHYSEKPM NWQRARRFCR DNYTDLVAIQ NKAEIHYLEK TLPFSRSYYW   50
Consensus    WTYHYSTKAM .W..AR.YCQ .RYTDLVAIQ NK.EIHYLNK .LPYS.SYYW   50

E-SELECTIN   IGIRKVNNVW MWVGTQKHLT HEAKNWAPGE PNNRQKDEDC VEIYIKREKD  100
P-SELECTIN   IGIRKNNKTW TWVGTKKALT NEAENWADNE PNNKRNNEDC VEIYIKSPSA  100
L-SELECTIN   IGIRKIGGIW TWVGTNKSLT HEAENWGDGE PNNKNKEDC VEIYIKRNKD  100
Consensus    IGIRK.N..W TWVGT.K.LT HEAENWADGE PNNK.N.EDC VEIYIKR.KD  100

E-SELECTIN   VGMWNDERCS KKKLALCYTA ACTNTSCSGH GECVETINNY TCKCDFGFSG  150
P-SELECTIN   FGKWNDEHCL KKKHALCYTA SCQDMSCSKQ GECLETIGNY TCSCYFGFYG  150
L-SELECTIN   AGKWNDDACH KLKAALCYTA SCQPWSCSGH GECVETINNY TCNCDVGYYG  150
Consensus    .GKWNDE.C. KKN.ALCYTA SCQ...SCSGH GECVETINNY TC.CDFGFYG  150
```

Figure 1. *Sequence homologies of the human selectin family.*

The high degree of homology found between the selectins strongly suggests that they were produced by duplication of an ancestral gene, followed by exon diversification and duplication. This is supported by the fact that the genes for all three proteins are clustered over a short region of human and mouse chromosome 1, indicating that gene duplication was prior to the evolutionary divergence of mouse and man.

STRUCTURE/FUNCTION RELATIONSHIPS

The fact that the selectins contain a region which is clearly related to a domain with known lectin activity, pointed unambiguously to the possibility that this was the portion of the molecules directly involved in cell binding and that the counter-receptors would be, at least in part, carbohydrate. A number of studies have shown that monoclonal antibodies which block selectin mediated adhesion, map to the lectin domain, confirming the importance of this region [5-7]. The role of the EGF-domain is, however, uncertain since it has been shown for E-selectin that deletion of the EGF-domain abolishes cell adhesion [6]. One possible role of the domain may be to maintain the structural conformation of the lectin domain. There is, however, evidence to support a direct role in cell binding, possibly by recognition of a protein component of the ligand [5,8].

The role of the CRP repeats, which vary in number from two in L-selectin, to nine in P-selectin, is also unproven. It is possible that they serve to optimize interactions with the counter-receptor, both by extending the lectin/EGF-domain away from the cell surface and by increasing molecular flexibility. However, a specific role in complement regulation cannot be excluded and would not be inappropriate at sites of inflammation.

LIGANDS

Although significant progress has been made towards identifying the counter-receptors for the selectins, much is still to be learnt. One of the paradoxes to emerge is that although the counter-receptor and in the case of L-selectin, the selectin itself, may be widely expressed they are none the less able to mediate highly specific cell–cell interactions. How this is achieved is far from clear, but a number of possibilities are emerging.

It appears that each member of the selectin family may recognize not one, but a number of ligands, which show differential cellular distribution and may have different binding characteristics, depending on the context in which they are presented. For example the sialyl Lewis^x (sLe^x) determinant is a major neutrophil ligand for E-selectin [9]. sLe^x, is however, weakly expressed, absent or masked on T cells, yet a subpopulation of T cells binds to E-selectin [10]. The T cell ligand is thought to be the cutaneous lymphocyte antigen [11]. Two of the molecules on neutrophils which carry the sLe^x determinant are of particular interest. Both E- and P-selectins can bind neutrophils via sLe^x present on L-selectin [12]. However, it is clear that this is not the only ligand on neutrophils since complete removal of L-selectin only reduces adhesion by 70%. Conversely, E-selectin cannot be the sole endothelial ligand for L-selectin, since a second counter-receptor shows a different time-course of expression [14].

A second way in which specificity may be controlled is by induced alterations in affinity. There are at least two reports to support such a mechanism. It has been shown that upon activation the capacity of neutrophils to bind E-selectin is decreased, although the number of sLe^x determinants remains constant [15]. A second example of affinity modulation is found with L-selectin. Activation of lymphocytes or neutrophils induces a rapid increase in the affinity of L-selectin for a polyphosphomannan ester, which bind specifically to L-selectin. In the case of L-selectin this effect could be induced by stimuli which were cell-type-specific, demonstrating the potential for producing a cell-type-specific interaction via a widely expressed receptor ligand pair [16].

Variations in post-translational modifications of the selectins may also effect specificity. For example, it has been shown that lymphocytes have L-selectin of a lower molecular weight than that of neutrophils and monocytes [17], although the functional significance of this is unknown.

The requirement for the simultaneous recognition of two independent determinants would serve to increase the specificity of selectin interactions. The possibility that such a mechanism may exist is suggested by studies with L-selectin showing that epitopes within the EGF-domain may be directly involved in cell binding, in addition to those in the lectin domain [5,8].

Finally, it is possible that an initial, selectin mediated interaction, common to many leukocytes, may have specificity introduced at another level. This is possible because leukocyte extravasation is a multistep process, in which an initial reversible interaction is followed by firm attachment mediated by the interaction of integrins and members of the immunoglobulin superfamily [18]. A cell-type-specific activation signal could, therefore, result in the adhesion of subsets of leukocytes being consolidated, whilst others simply detach and re-enter the circulation. The activation step is dependent upon a signal delivered to the leukocyte, which triggers an increase in the affinity of the leukocyte integrins for the endothelial ligand. Such a signal may be either delivered by a soluble activator, such as platelet activating factor, or may be contact mediated. Interestingly, there is evidence that E-selectin is capable of delivering such a signal to neutrophils. [19].

FUNCTION

Recent studies suggest that the selectins are involved in the earliest phases of a cascade of events leading to leukocyte extravasation. It is a long-standing observation that in regions of inflammation, leukocytes move towards the edge of the capillary and begin to roll along the endothelium. The involvement of P-selectin in this process has been demonstrated *in vitro* [20]. *In vivo* studies with an antibody to L-selectin suggest that it too is involved in rolling [21]. A function with E-selectin in leukocyte rolling has yet to be demonstrated, but seems probable.

The process of rolling requires that an interaction takes place between a receptor on the endothelial cell and a counter-receptor on a moving leukocyte and that this interaction is then broken and the cycle repeated. Under such conditions, when cells are moving rapidly in relation to one another, the rate of reaction may be of greater importance than the absolute affinity. The structure of the selectins may be uniquely suited to mediating rapid interactions. It has been argued that the presence of multiple consensus repeats in the selectins allows for a high degree of flexibility in the molecule [20]. Carbohydrate determinants of the ligand may also be presented on flexible chains. This combination of flexibilities would allow the selectin and carbohydrate to rapidly adopt the correct conformation for binding to take place [22]. The rate of reaction would be further enhanced if the carbohydrate determinants were present at high densities.

CONCLUSIONS

The enormous potential of carbohydrates for encoding specific information which could be used in cell recognition has long been appreciated. Until recently their role in vertebrates has been largely of theoretical, rather than practical importance. The discovery of the selectin family has placed a major role in transforming this image. Although rapid progress has been made in identifying the minimal structures for recognition, much remains to be learnt about the context in which they are presented to confer specificity and how selectin mediated interactions are interlinked with other adhesion pathways.

References
1 Vestweber, D. (1992) Seminars Cell Biol. 3, 211–220.
2 Lasky, A.L. (1992) Science 258, 964–969.
3 Drickamer, K. (1989) Ciba Found. Symp. 145, 45–61.
4 Disdier, M. et al. (1991) J. Cell Biol. 114, Abstract 1754.
5 Kansas, S.K. et al. (1991) J. Cell Biol. 114, 351–358.
6 Pigott, R. et al. (1991) J. Immunol. 147, 130–135.
7 McEver, R.P. (1991) In Cellular and Molecular Mechanisms of Inflammation, (Cochrane, C.G. and Gimbrone, M. A. eds), pp.15–29. Academic Press, New York.
8 Seigelman, M.H. et al. (1990) Cell 61, 611–622.
9 Walz, G. et al. (1990) Science 250, 1132–1135.
10 Picker, L.J. et al. (1991) Nature 349, 796–797.
11 Berg, E.L. et al. (1991) J. Exp. Med. 174, 1461–1466.
12 Picker, L.J. et al. (1991). Cell 66, 921–933.
13 Gahmberg, C. et al. Glycobiology, in press.
14 Spertini, O. et al. (1991) J. Immunol. 147, 2565–2573.
15 Gimbrone, M. et al. (1989) Science 246, 1601–1603.
16 Spertini, O. et al. (1991) Nature 349, 691–693.
17 Lweinsohn, D.M. et al. (1987) J. Immunol. 138, 4313–4321.
18 Springer, T.A. (1990) Nature 346, 425–434.
19 Lo, S.K. (1991) J. Exp. Med. 173, 1493–1500.
20 Lawrence, M.B. and Springer, T.A. (1991) Cell 65, 859–873.
21 von Adrian, U.H. et al. (1991) Proc. Natl Acad. Sci. USA 87, 7538–7542.
22 Williams, A.F. (1991) Nature, 352, 473–474.

Section II

THE ADHESION MOLECULES

AMOG — adhesion molecule on glia

Family
AMOG is an isoform of the β subunit of the Na,K-ATPase [1].

Cellular distribution
CNS astrocytes [2].

Function
AMOG mediates the adhesion of neurons and astrocytes *in vitro* and has a role in neuronal migration [3]. Adhesion is calcium-independent. AMOG is associated with the catalytic α subunit of the Na,K-ATPase.

Ligand
Unknown.

Gene location and size
11; <7.2 kb [4].

Molecular weights
Polypeptide   33 287
SDS PAGE   45 000–50 000

Amino acid sequence [1]

```
  1  MVIQKEKKSC GQVVEEWKEF VWNPRTHQFM GRTGTSWAFI LLFYLVFYGF
                                                          *
 51  LTAMFSLTMW VMLQTVSDHT PKYQDRLATP GLMIRPKTEN LDVIVNISDT
                        *
101  ESWGQHVQKL NKFLEPYNDS IQAQKNDVCV PGRYYEQPDN GVLNYPKRAC
                                                 *        *
151  QFNRTQLGDC SGIGDPTHYG YSTGQPCVFI KMNRVINFYA GANQSMNVTC
                                  *              *
201  VGKRDEDAEN LGHFVMFPAN GSIDLMYFPY YGKKFHVNYT QPLVAVKFLN
251  VTPNVEVNVE CRINAANIAT DDERDKFAGR VAFKLRINKT
```

Domain structure
1–39   Cytoplasmic domain
40–67   Transmembrane domain
68–290 Extracellular domain

Database accession numbers
	PIR	SWISSPROT	EMBL/GENBANK
Mouse	A34057	P14231	X56007

References
1 Gloor, S. et al. (1990) J. Cell Biol. 110, 165–174.
2 Paliusi, S.R. et al. (1990) Eur.J. Neurosci. 2, 471–580.
3 Antonicek, H. et al. (1987) J. Cell Biol. 104, 1587–1595.
4 Magyar, J.P. and Schachner, M. (1990) Nucleic Acids Res. 18, 6995–6996.

B-Cadherin K-CAM*

Family
Cadherin.

Cellular distribution
Found in chick embryonic tissue, especially the epithelial lining of the choroid plexus and the optic tectum of chick brain [1]. Also expressed in diverse non-neural tissues such as liver, eye, heart, intestine, bladder, skeletal muscle, skin and retina. The reported distribution of K-CAM is not identical to that of B-cadherin [2].

Function
May play a role in Ca^{2+} dependent cell adhesion. May be important in organizing calcium-dependent junctional complexes which function as selective filters between cerebrospinal fluid and brain. May also play a role in the formation of discrete neural cell layers [1].

Regulation of expression
Developmentally regulated. Protein expression gradually increases from embryonic day E6 to E16 then declines [1].

Ligand
Unknown. May mediate homophilic adhesion.

Gene structure

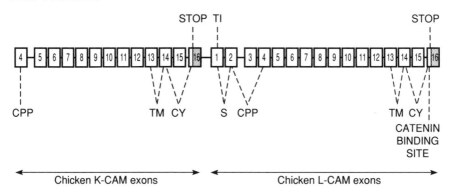

K-CAM gene is arranged in tandem with the chicken L-CAM gene, both spanning 15 kb [2]. The K-CAM gene (exons 4–15) spans approximately 4 kb and is separated from the chicken L-CAM gene by only 700 bp. The sizes of exons 4–15 are almost identical to those found in chicken L-CAM except for exon 13 (one codon shorter in K-CAM). Intron/exon boundaries are also the same for the two genes. Exon 16 in K-CAM (3' untranslated region) is approximately 600 bp shorter than in L-CAM. Intron sizes and sequences are not conserved between the two genes.

Both genes may share a promoter or both mRNAs may be generated from the same initial transcript by alternative splicing [2].

B-Cadherin

Molecular weights
Polypeptide   79 742
SDS PAGE   120 000 and 122 000 forms

Amino acid sequence (from embryonic day 13 chick brain) [1]

```
  1  DWVIPPIKVP  ENERGPFPKN  LVQIKSNRDR  EAKIFYSITG  QGADAPPEGI

 51  FTIEKETGWM  KVTQPLDREH  INKYHLYSHA  VSENGKPVEE  PMEIIVTVTD
                                              *
101  QNDNKPQFTQ  EVFRGSVPEG  ALPGTSVMRV  NATDADDDVE  TYNGVIAYSI

151  LSQEPREPHP  HMFTVNRATG  TLSVIASGLD  RERVREYTLT  MQAADLDGQG

201  LTTTALAVIE  ITDVNDNAPE  FDPKTYEAAV  PENEAELEVA  RLATTDLDEP

251  HTPAWRAVYS  IVRGNEGGAF  TITTDPASNE  GVLRTAKGLD  YEAKRQFVLH

301  VAVVNEAPFA  IKLPTATATV  MVSVEDVNEA  PVFDPPLRLA  QVPEDVPLGQ

351  PLASYTAQDP  DRAQQQRIKY  VMGSDPAGWL  AVHPENGLIT  AREQLDRESP
          *                                                  *
401  FTKNSTYVAV  LLAVDDGLPP  ATGTGTLLLT  LLDVNDHGPE  PEPRDIVICN

451  RSPVPQVLTI  TDRDLPPNTG  PFRAELSHGS  GDSWAVEVGN  GGDTVALWLT

501  EPLEQNLYSV  YLRLFDRQGK  DQVTVIRAQV  CDCQGRVESC  AQKPRVDTGV

551  PIVLAVGAV   LALLLVLLLL  LLLVRRRKVV  KEPLLLPEDD  TRDNIFYYGE

601  EGGGEEDQDY  DLSQLHRGLD  ARPEVIRNDV  APPLMAAPQY  RPRPANPDEI

651  GNFIDENLKA  ADTDPTAPPY  DSLLVFDYEG  GGSEATSLSS  LNSSASDQDQ

701  DYDYLNEWGN  RFKKLAELYG  GGEDEE
```

Domain structure

The extracellular domain contains five internal repeats (EC1–EC5) of approximately 112 amino acids.

The putative calcium binding motifs are within EC1, EC2 and EC3 (residues 100–104, 134–136 and 213–217 respectively).

    1–548   Extracellular domain
    1–108   EC1
  109–221   EC2
  222–333   EC3
  334–449   EC4
  450–548   EC5
  549–574   Transmembrane domain
  575–726   Cytoplasmic domain

Three potential N-linked glycosylation sites.

B-Cadherin

Database accession numbers

	PIR	SWISSPROT	EMBL/GENBANK	REFERENCE
Chicken			X58518	1
			M81894	2

Alternative forms

B-Cadherin migrates as a doublet of 120 and 122 kD on SDS PAGE. The 122 kD band may represent a precursor or post-translationally modified form.

Chicken K-CAM*[2] may be a polymorphic form of B-cadherin (due to species differences). It differs by 12 bp from B-cadherin (representing only a single amino acid change – Met_{408} instead of Val) and is detectable in kidney whereas B-cadherin is not.

References
[1] **Napolitano, E.W. et al. (1991) J. Cell Biol. 113, 893–905.**
[2] Sorkin, B.C. et al. (1991) Proc. Natl Acad. Sci. USA 88, 11545–11549.

CD2

LFA-2, T11, leu-5, Tp50

Family
Immunoglobulin superfamily. CD2 is structurally similar to its ligand LFA-3, although they share only 21% amino acid identity.

Cellular distribution
Thymocytes, T lymphocytes, NK cells.

Function
CD2 is an accessory molecule important in mediating the interaction of activated T cells and thymocytes with antigen-presenting cells and target cells. Signal transduction involving CD2 is important in T cell stimulation [reviewed in ref.1]. Specific combinations of anti-CD2 antibodies, or a single antibody in combination with LFA-3, are able to stimulate T cells. There is evidence that there may be a cooperative effect between CD2 and the T cell receptor [reviewed in ref.1] since signal transduction in cells lacking the TCR is poor. There is evidence that CD2 may trigger LFA-3 [2].

Ligands
LFA-3 (CD58) [3], CD59 [4].

Gene Structure [5]

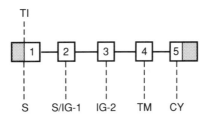

Gene location and size
1p13; <28.5 kb.

Structure [6]

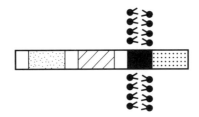

The crystal structure of CD2 has been reported [6]. Domain 1 resembles an immunoglobulin V domain, although it lacks the disulphides characteristic of most Ig-domains. The LFA-3 and CD59 binding sites are within domain 1 and are overlapping but non-identical [4].

CD2

Molecular weights
Polypeptide 36 885
SDS Page 47 000–58 000

Amino acid sequence [7]

```
                    ↓
  1  MSFPCKFVAS FLLIFNVSSK GAVSKEITNA LETWGALGQD INLDIPSFQM
                                           *
 51  SDDIDDIKWE KTSDKKKIAQ FRKEKETFKE KDTYKLFKNG TLKIKHLKTD
                                           *              *
101  DQDIYKVSIY DTKGKNVLEK IFDLKIQERV SKPKISWTCI NTTLTCEVMN
151  GTDPELNLYQ DGKHLKLSQR VITHKWTTSL SAKFKCTAGN KVSKESSVEP
201  VSCPEKGLDI YLIIGICGGG SLLMVFVALL VFYITKRKKQ RSRRNDEELE
251  TRAHRVATEE RGRKPQQIPA STPQNPATSO HPPPPPGHRS OAPSHRPPPP
301  GHRVQHQPQK RPPAPSGTQV HQQKGPPLPR PRVQPKPPHG AAENSLSPSS
351  N
```

Domain structure
1–24 Signal sequence
41–111 Ig-1
142–190 Ig-2
210–235 Transmembrane domain

Three potential N-linked glycosylation sites.

Database accession numbers

	PIR	SWISSPROT	EMBL/GENBANK	REFERENCE
Human	A28967	P06729	M16445	
Mouse	S02293	P08920	Y00023	[8]
Rat	A33071	P08921	X05111	[9]

Alternative forms
None known.

References
[1] **Dustin, M.L. and Springer, T.A. (1990) Annu. Rev. Immunol. 9, 27–66.**
[2] Hahn, W.C. et al. (1991) J. Immunol. 147, 14–21.
[3] Selveraj, P. et al. (1987) Nature 326, 400–403.
[4] Hahn, W.C. et al. (1992) Science 256, 1805–1807.
[5] Lang, G. et al. (1988) EMBO. J. 7, 1675–1682.
[6] Jones, Y.E. et al. (1992) Nature 360, 232–239.
[7] Seed, B. and Aruffo, A. (1987) Proc. Natl Acad. Sci. USA 84, 3365–3369.
[8] Sewell, W.A. et al. (1987) Eur. J. Immunol. 17, 1015–1020.
[9] Williams, A.F. et al. (1987) J. Exp. Med. 165, 368–380.

CD4

T4, L3T4 (mouse), W3/25 (rat)

Family
Immunoglobulin superfamily.

Cellular distribution
Thymocytes, T lymphocytes (primarily T helper), monocytes, dendritic cells.

Function
CD4 is a co-receptor with the T cell receptor involved in antigen recognition. CD4 enhances T cell activation by increasing the avidity of T cells for effector and target cells and by signal transduction. The increase in avidity is probably most important in circumstances where the affinity of the T cell receptor for antigen is low, or when antigen is limiting [reviewed in ref. 1]. CD4 is a receptor for HIV-1. HIV-1 gp120 binds within the N-terminal domain [2,3]. The cytoplasmic domain of CD4 is implicated in signal transduction. Deletion of the cytoplasmic tail can prevent T cell stimulation without effecting cell adhesion [4]. CD4 is associated with the tyrosine kinase p56[lck] [reviewed in ref. 5]. CD4 is important in the selection process by which CD4-positive T cells become MHC Class II restricted [6].

Ligand
MHC Class II [7].

Gene Structure [8]

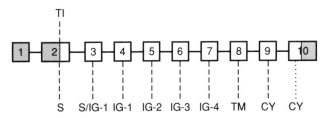

Gene location and size
12pter-p12; 33 kb.

Structure

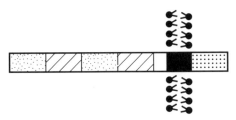

Domains 1 and 2 have been confirmed as immunoglobulin-like, although domain 2 contains an unusual disulphide link [2,3]. Domain 3 lacks one of the disulphides characteristic of most Ig-domains.

CD4

Molecular weights
Polypeptide   48 398
SDS Page   55 000

Amino acid sequence [9]

```
                            ↓
  1   MNRGVPFRHL LLVLQLALLP AATQGKKVVL GKKGDTVELT CTASQKKSIQ

 51   FHWKNSNQIK ILGNQGSFLT KGPSKLNDRA DSRRSLWDQG NFPLIIKNLK

101   IEDSDTYICE VEDQKEEVQL LVFGLTANSD THLLQGQSLT LTLESPPGSS

151   PSVQCRSPRG KNIQGGKTLS VSQLELQDSG TWTCTVLQNQ KKVEFKIDIV

201   VLAFQKASSI VYKKEGEQVE FSFPLAFTVE KLTGSGELWW QAERASSSKS
                                                          *
251   WITFDLKNKE VSVKRVTQDP KLQMGKKLPL HLTLPQALPQ YAGSGNLTLA
                                    *
301   LEAKTGKLHQ ENVLVVMRAT QLQKNLTCEV WGPTSPKLML SLKLENKEAK

351   VSKREKAVWV LNPEAGMWQC LLSDSGQVLL ESNIKVLPTW STPVQPMALI

401   VLGGVAGLLL FIGLGIFFCV RCRHRRRQAE RMSQIKRLLS EKKTCQCPHR

451   FQKTCSPI
```

Domain structure
   1–25   Signal sequence
  26–125   Ig-1
 126–203   Ig-2
 204–317   Ig-3
 318–374   Ig-4
 397–418   Transmembrane domain

Database accession numbers
	PIR	SWISSPROT	EMBL/GENBANK	REFERENCE
Human	A02109	P01730	M12807	
Mouse	A26038	P06332	M13816	*10*
Rat	A27449	P05540	M15768	*11*

Alternative forms
None known.

References

1 **Parnes, J.R. et al. (1989) Cold Spring Harb. Symp. Quant. Biol. 54, 649–655.**
2 Wang, J. et al. (1990) Nature 348, 411–418.
3 Ryu, S.-E. et al. (1990) Nature 348, 419–425.
4 Sleckman, B.P. et al. (1991) J. Immunol. 147, 428–431.
5 Rudd, C.E. (1990) Immunol. Today 11, 400–406.
6 Ramsdell, F. and Fowlkes, B.J. (1989) Cell, 99–107.
7 Doyle, C. and Strominger, J.L. (1987) Nature 330, 256–259.
8 Maddon, P.J. et al. (1987) Proc. Natl Acad. Sci. USA 84, 9155–9159.
9 Maddon, P.J. et al. (1985) Cell 42, 93–104.
10 Tourvieille, B. et al. (1986) Science 234, 610–614.
11 Clark, S.J. et al. (1987) Proc. Natl Acad. Sci. USA 84, 1649–1653.

CD8

Leu2/T8, Lyt2/Lyt3 (mouse), OX-8 (rat)

Family
Immunoglobulin superfamily.

Cellular distribution
Thymocytes, T cells (primarily cytotoxic and suppressor), natural killer (NK) cells.

Function
CD8 is involved in antigen recognition. Binding of CD8 to its ligand, MHC Class I, on target and effector cells, enhances T cell activation [reviewed in ref.1] The cytoplasmic domain of CD8 is associated with tyrosine kinase p56[lck] and is involved in signal transduction [reviewed in ref.2]. CD8 is important in the process by which CD8-positive T cells become MHC Class I restricted [3].

Ligand
Non-polymorphic regions of MHC Class I [4].

Gene Structure (CD8α) [5]

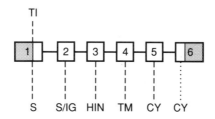

Gene location and size
2p12; 7 kb (CD8α). 2; >15 kb (CD8β).

Structure

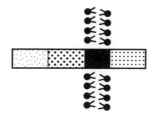

CD8 consists of heterodimers of CD8α/CD8β and possibly also homodimers of CD8α [6]. The crystal structure for CD8α homodimers has been reported [7].

Molecular weights
Polypeptide   23 551 (CD8α); 21 351 (CD8β)
Reduced SDS Page   32 000–34 000 (CD8α and β)

CD8

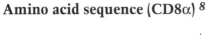

Amino acid sequence (CD8α) [8]

```
                          ↓                                    *
  1  MALPVTALLL PLALLLHAAR PSQFRVSPLD RTWNLGETVE LKCQVLLSNP
 51  TSGCSWLFQP RGAAASPTFL LYLSQNKPKA AEGLDTQRFS GKRLGDTFVL
101  TLSDFRRENE GYYFCSALSN SIMYFSHFVP VFLPAKPTTT PAPRPPTPAP
151  TIASQPLSLR PEACRPAAGG AVHTRGLDFA CDIYIWAPLA GTCGVLLLSL
201  VITLYCNHRN RRRVCKCPRP VVKSGDKPSL SARYV
```

Domain structure

1–21 Signal sequence
36–120 Ig
183–208 Transmembrane domain

Amino acid sequence (CD8β-1) [6]

```
                          ↓
  1  MRPRLWLLLA AQLTVLHGNS VLQQTPAYIK VQTNKMVMLS CEAKISLSNM
 51  RIYWLRQRQA PSSDSHHEFL ALWDSAKGTI HGEEVEQEKI AVFRDASRFI
         *
101  LNLTSVKPED SGIYFCMIVG SPELTFGKGT QLSVVDFLPT TAQPTKKSTL
151  KKRVCRLPRP ETQKGPLCSP ITLGLLVAGV LVLLVSLGVA IHLCCRRRRA
201  RLRFMKQFYK
```

Domain structure

1–21 Signal sequence
22–123 Ig
165–195 Transmembrane domain

Database accession numbers

	PIR	SWISSPROT	EMBL/GENBANK	REFERENCE
Human CD8α	A01999	P01732	M27161	
Human CD8β-1	S01649	P10966	X13444	
Human CD8β-2	S01873	P14860	X13445	
Human CD8β-3	S01874	P14861	X13446	
Mouse CD8α	A01998	P01731	M12825	9
Mouse CD8β	S03408	P10300	X07698	11
Rat CD8α	A24637	P07725	X03015	10
Rat CD8β	A24184	P05541	X04310	12

Alternative forms

Alternative splicing of CD8α produces a soluble variant [5]. A soluble form of CD8, CD8β3, has also been predicted, in addition to two variants differing in the length of the cytoplasmic domain [6]. An alternatively spliced variant of murine CD8α, CD8α' has a shortened cytoplasmic domain [13].

References
1. Parnes, J.R. el al. (1989) Cold Spring Harb. Symp. Quant. Biol. 54, 649–655.
2. Rudd, C.E. (1990) Immunol. Today 11, 400–406.
3. Robey, E.A. et al. (1991) Cell 65, 99–107.
4. Salter, R.D.. et al. (1990) Nature 345, 41–46.
5. Norment, A.M. and Littman D.R. (1989) J. Immunol. 142, 3312–3319.
6. Norment, A.M. and Littman D.R. (1988) EMBO. J. 7, 3433–3439.
7. Leahy, D.J. et al. (1992) Cell 68, 1145–1162.
8. Littman, D.R. et al. (1985) Cell 40, 237–246.
9. Nakauchi, H. et al. (1985) Proc. Natl Acad. Sci. USA. 82, 5126–5130.
10. Johnson, P. et al. (1985) EMBO. J. 4, 2539–2545.
11. Blanc, D. et al. (1985) Eur. J. Immunol. 18, 613–619.
12. Johnson, P. and Williams, A.F. (1986) Nature 323, 74–76.
13. Zamoyska, R. et al. (1985) Cell, 153–163.

CD22

BL-CAM (B lymphocyte cell adhesion molecule)

Family
Immunoglobulin superfamily.

Cellular distribution
Subset of B-lymphocytes.

Function
CD22 is thought to have a role in mediating antigen non-specific interactions of B cells, leading to activation. CD22α is involved in the adhesion of B cells to monocytes and erythrocytes [1]. CD22β mediates both B cell/B cell interactions [2] and B cell/T cell interactions [3]. Multiple potential phosphorylation sites in the cytoplasmic domain suggest a role in signalling. Anti-CD22 antibodies augment entry into cell cycle and increase anti-Ig-induced rises in intrathymic calcium [5].

Regulation of expression
CD22 is present in the cytoplasm of pre-B-cells, surface expression parallels IgM expression and is lost upon differentiation [4].

Ligands
CD22β binds sialylated glycoproteins, one of which, CD45, a receptor-linked phosphotyrosine phosphotase is reported to be a T cell ligand [3]. In addition, CD22β binds to glycoproteins and possibly glycolipids, sialylated by α2,6-sialyltransferase on both T and B cells.

Structure (CD22β)

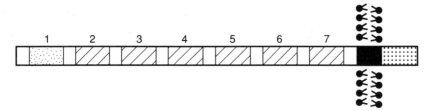

CD22 is a heterodimer of two independently expressed polypeptide chains, CD22α and CD22β. CD22α lacks immunoglobulin-like domains 3 and 4. The T and B cell binding sites are within the three N-terminal domains and involve distinct epitopes [3].

Molecular weights
Polypeptide   70 991 (CD22α); 95 000 (CD22β)
SDS PAGE   130 000 (CD22α); 140 000 (CD22β)

CD22

Amino acid sequence [1]

```
                         ↓
  1  MHLLGPWLLL  LVLEYLAFSD  SSKWVFEHPE  TLYAWEGACV  WIPCTYRALD
                      *
 51  GDLESFILFH  NPEYNKNTSK  FDGTRLYEST  KDGKVPSEQK  RVQFLGDKNK
          *           *                       *
101  NCTLSIHPVH  LNDSGQLGLR  MESKTEKWME  RIHLNVSERP  FPPHIQLPPE
                                  *
151  IQESQEVTLT  CLLNFSCYGY  PIQLQWLLEG  VPMRQAAVTS  TSLTIKSVFT
                                              *
201  RSELKFSPQW  SHHGKIVTCQ  LQDADGKFLS  NDTVQLNVKH  PPKKVTTVIQ
                     *
251  NPMPIREGDT  VTLSCNYNSS  NPSVTRYEWK  PHGAWEEPSL  GVLKIQNVGW
          *
301  DNTTIACAAC  NSWCSWASPV  ALNVQYAPRD  VRVRKIKPLS  EIHSGNSVSL
                                                            *
351  QCDFSSSHPK  EVQFFWEKNG  RLLGKESQLN  FDSISPEDAG  SYSCWVNNSI
401  GQTASKAWTL  EVLYAPRRLR  VSMSPGDQVM  EGKSATLTCE  SDANPPVSHY
          *
451  TWFDWNNQSL  PYHSQKLRLE  PVKVQHSGAY  WCQGTNSVGK  GRSPLSTLTV
501  YYSPETIGRR  VAVGLGSCLA  ILILAICGLK  LQRRWKRTQS  QQGLQENSSG
551  QSFFVRNKKV  RRAPLSEGPH  SLGCYNPMME  DGISYTTLRF  PEMNIPRTGD
601  AESSEMQRPP  PDCDDTVTYS  ALHKRQVGTM  RTSFQIFQKM  RGFITQS
```

Domain structure
  1–19   Signal sequence
  37–106 Ig-1
  154–223 Ig-2
  258–311 Ig-3
  345–398 Ig-4
  432–486 Ig-5
  511–529 Transmembrane domain

Ten potential N-linked glycosylation sites.

Database accession numbers
	PIR	SWISSPROT	EMBL/GENBANK	REFERENCE
Human CD22α	A35648	P20273		
Mouse CD22β	The sequence is not yet available in the databases			2

References
[1] Stamenkovic, I. and Seed, B. (1990) Nature 345, 74–77.
[2] Wilson, G.L. et al. (1991) J. Exp. Med. 173, 137–146.
[3] **Stamenkovic, I. et al. (1991) Cell 66, 1133–1144.**
[4] Dorken, B. et al. (1986) J. Immunol. 136, 4470–4479.
[5] Pezzutto, A. et al. (1988) J. Immunol. 140, 1791–1795.

CD23 — FcεRII, low affinity IgE receptor

Family
Ca^{2+} dependent C-type lectin.

Cellular distribution

Follicular dendritic cells, B lymphocytes, T lymphocytes, eosinophils, monocytes, macrophages, platelets, a subset of thymic epithelial cells, Langerhans cells, NK cells [1].

Function

Regulation of IgE production. Differentiation and proliferation of B cells. Also possibly involved in antigen presentation to T cells [1,2]. In the presence of IL1, CD23 is proposed to be involved in the rescue of germinal centre B cells from apoptosis.

Regulation of expression

IL4 is the major upregulator of CD23 and its soluble form (25 kD) on normal B cells, human Burkitt lymphoma-derived cell lines, U937 cells and the eosino-philic cell line EoLs. ECF-A, PAF, IL5, GM-CSF, TNFα can also upregulate CD23. EBV is an upregulator of CD23; EBNA2 transactivates CD23a; LMP transactivates CD23b (see below).
PGE$_2$, dexamethasone, IFNα, IFNγ and TGFβ have a suppressive effect on CD23 expression [1].

Ligands
IgE [1], CD21 [3].

Gene structure

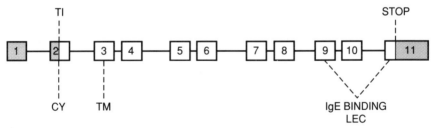

exon
1   Untranslated leader
2   Translation initiation
3   Transfer stop and membrane anchor
4–11   Extracellular domain
9–11   IgE binding domain/lectin homology
11   Contains stop codon and 3' untranslated region

The gene spans 13 kb and contains 11 exons. A 188 bp long inverted repeat flanks the promoter and may be involved in gene regulation [4].

CD23

Gene location and size
19[5]; 13 kb.

Molecular weights
Polypeptide 33 888
SDS PAGE 45 000 (membrane-bound form). A soluble form of 25 000 arises by proteolytic cleavage of membrane bound CD23 (amino acids 150–321). Additional soluble forms of MW 37 000, 33 000 and 17 000 have also been identified [1].

Amino acid sequence (from human B lymphoblastoid cell line RPMI 8866) [6]

```
  1  MEEGQYSEIE ELPRRRCCRR GTQIVLLGLV TAALWAGLLT LLLLWHWDTT
                          *
 51  QSLKQLEERA ARNVSQVSKN LESHHGDQMA QKSQSTQISQ ELEELRAEQQ
101  RLKSQDLELS WNLNGLQADL SSFKSQELNE RNEASDLLER LREEVTKLRM
151  ELQVSSGFVC NTCPEKWINF QRKCYYFGKG TKQWVHARYA CDDMEGQLVS
201  IHSPEEQDFL TKHASHTGSW IGLRNLDLKG EFIWVDGSHV DYSNWAPGEP
251  TSRSQGEDCV MMRGSGRWND AFCDRKLGAW VCDRLATCTP PASEGSAESM
301  GPDSRPDPDG RLPTPSAPLH S
```

Domain structure
 1–21   Cytoplasmic NH$_2$-terminal
22–47   Transmembrane domain
48–321   Extracellular domain
171–283 Lectin domain

One potential N-linked glycosylation site.

Database accession numbers
	PIR	SWISSPROT	EMBL/GENBANK	REFERENCE
Human	A26164	P06734	M15059	6
Mouse	A33840	P20693		
			M34163	7

Alternative forms
CD23a and CD23b result from alternative splicing at the N-terminal and they differ in the 5' untranslated region and in the cytoplasmic domain [8]. CD23a is restricted to B cells.

References

1. Fridman, W.H. (ed.) (1989) Structures and Functions of Low Affinity Fc Receptors, Chemical Immunology Vol.47, Karger, Basel.
2. **Delespesse, G. et al. (1992) Immunol. Rev. 125, 77–97.**
3. Aubry J-P. et al. (1992) Nature 358, 505–507.
4. Suter, U. et al. (1987) Nucleic Acids Res. 15, 7295–7308.
5. Wendel Hansen, V. et al. (1990) Somat. Cell. Mol. Genet. 16, 283.
6. Ikuta, K. et al. (1987) Proc. Natl Acad. Sci. USA. 84, 819–823.
7. Gollnick, S.O. et al. (1990) J. Immunol. 144, 1974–1982.
8. Stengelin, S. et al. (1988) EMBO J. 7, 1053–1059.

CD35 — Complement receptor type I (CR1), C3b/C4b receptor

Family
Short consensus repeat (SCR) family.

Cellular distribution
Erythrocytes, leukocytes, glomerular podocytes, splenic follicular dendritic cells.

Function
Binding of particles and immune complexes that have activated complement. Reversibly binds the C3b and C4b fragments of C3 and C4 which are covalently attached to complement activating surfaces. Erythrocyte CD35 binds immune complexes for transport to liver. CD35 on neutrophils and monocytes internalizes bound complexes (either by absorptive endocytosis through coated pits or by phagocytosis). Phagocytosis requires activation of the receptor by phorbol esters, chemotactic peptides or ECM proteins probably through phosphorylation of CD35. On dendritic cells, CD35 may have a role in antigen presentation [1,2].

CD35 inhibits complement activation by two distinct mechanisms. CD35 is a cofactor for factor I cleavage of C3b, iC3b and C4b. CD35 mediates decay-dissociation of C3 and C5 convertases [3].

Regulation of expression
Activation of PMNs *in vitro* by GM-CSF leads to increased surface expression of CD35 [4]. In resting neutrophils, CD35 is stored in empty looking vesicles which are distinct from typical granules [5].

Ligands
C3b and C4b. Two binding sites for C3b are located in SCR domains 8–11 and 15–18. The binding site for C4b is located in SCR 1–4 [6–8].

Gene structure
The gene encoding the B or S (which contains five LHRs) allele spans 160 kb and contains 42 exons. Distinct exons encode the signal peptide, transmembrane domain, cytoplasmic domain and most of the SCRs except the second SCR in each LHR [9].

Gene location
1q32 (RCA locus) [10, 11].

Molecular weights
Polypeptide 219 635 (F allotype)
SDS Page   235 000 (F allotype); 290 000 (S allotype)

CD35

Amino acid sequence (from overlapping clones from a tonsil cDNA library and a DMSO-induced HL-60 cDNA library) [1,6]

↓

```
   1  MGASSPRSPE PVGPPAPGLP FCCGGSLLAV VVLLALPVAW GQCNAPEWLP
        *
  51  FARPTNLTDE FEFPIGTYLN YECRPGYSGR PFSIICLKNS VWTGAKDRCR
 101  RKSCRNPPDP VNGMVHVIKG IQFGSQIKYS CTKGYRLIGS SSATCIISGD
 151  TVIWDNETPI CDRIPCGLPP TITNGDFIST NRENFHYGSV VTYRCNPGSG
 201  GRKVFELVGE PSIYCTSNDD QVGIWSGPAP QCIIPNKCTP PNVENGILVS
        *
 251  DNRSLFSLNE VVEFRCQPGF VMKGPRRVKC QALNKWEPEL PSCSRVCQPP
 301  PDVLHAERTQ RDKDNFSPGQ EVFYSCEPGY DLRGAASMRC TPQGDWSPAA
 351  PTCEVKSCDD FMGQLLNGRV LFPVNLQLGA KVDFVCDEGF QLKGSSASYC
        *                                            *
 401  VLAGMESLWN SSVPVCEQIF CPSPPVIPNG RHTGKPLEVF PFGKAVNYTC
 451  DPHPDRGTSF DLIGESTIRC TSDPQGNGVW SSPAPRCGIL GHCQAPDHFL
        *
 501  FAKLKTQTNA SDFPIGTSLK YECRPEYYGR PFSITCLDNL VWSSPKDVCK
                            *
 551  RKSCKTPPDP VNGMVHVITD IQVGSRINYS CTTGHRLIGH SSAECILSGN
 601  AAHWSTKPPI CQRIPCGLPP TIANGDFIST NRENFHYGSV VTYRCNPGSG
 651  GRKVFELVGE PSIYCTSNDD QVGIWSGPAP QCIIPNKCTP PNVENGILVS
        *
 701  DNRSLFSLNE VVEFRCQPGF VMKGPRRVKC QALNKWEPEL PSCSRVCQPP
 751  PDVLHAERTQ RDKDNFSPGQ EVFYSCEPGY DLRGAASMRC TPQGDWSPAA
 801  PTCEVKSCDD FMGQLLNGRV LFPVNLQLGA KVDFVCDEGF QLKGSSASYC
        *                                            *
 851  VLAGMESLWN SSVPVCEQIF CPSPPVIPNG RHTGKPLEVF PFGKAVNYTC
 901  DPHPDRGTSF DLIGESTIRC TSDPQGNGVW SSPAPRCGIL GHCQAPDHFL
        *
 951  FAKLKTQTNA SDFPIGTSLK YECRPEYYGR PFSITCLDNL VWSSPKDVCK
1001  RKSCKTPPDP VNGMVHVITD IQVGSRINYS CTTGHRLIGH SSAECILSGN
1051  TAHWSTKPPI CQRIPCGLPP TIANGDFIST NRENFHYGSV VTYRCNLGSR
```

CD35

```
1101    GRKVFELVGE  PSIYCTSNDD  QVGIWSGPAP  QCIIPNKCTP  PNVENGILVS
                *
1151    DNRSLFSLNE  VVEFRCQPGF  VMKGPRRVKC  QALNKWEPEL  PSCSRVCQPP
1201    PEILHGEHTP  SHQDNFSPGQ  EVFYSCEPGY  DLRGAASLHC  TPQGDWSPEA
1251    PRCAVKSCDD  FLGQLPHGRV  LFPLNLQLGA  KVSFVCDEGF  RLKGSSVSHC
                                *
1301    VLVGMRSLWN  NSVPVCEHIF  CPNPPAILNG  RHTGTPSGDI  PYGKEISYTC
1351    DPHPDRGMTF  NLIGESTIRC  TSDPHGNGVW  SSPAPRCELS  VRAGHCKTPE
1401    QFPFASPTIP  INDFEFPVGT  SLNYECRPGY  FGKMFSISCL  ENLVWSSVED
                                                            *
1451    NCRRKSCGPP  PEPFNGMVHI  NTDTQFGSTV  NYSCNEGFRL  IGSPSTTCLV
            *                               *      *
1501    SGNNVTWDKK  APICEIISCE  PPPTISNGDF  YSNNRTSFHN  GTVVTYQCHT
1551    GPDGEQLFEL  VGERSIYCTS  KDDQVGVWSS  PPPRCISTNK  CTAPEVENAI
                        *
1601    RVPGNRSFFS  LTEIIRFRCQ  PGFVMVGSHT  VQCQTNGRWG  PKLPHCSRVC
1651    QPPPEILHGE  HTLSHQDNFS  PGQEVFYSCE  PSYDLRGAAS  LHCTPQGDWS
1701    PEAPRCTVKS  CDDFLGQLPH  GRVLLPLNLQ  LGAKVSFVCD  EGFRLKGRSA
                            *
1751    SHCVLAGMKA  LWNSSVPVCE  QIFCPNPPAI  LNGRHTGTPF  GDIPYGKEIS
1801    YACDTHPDRG  MTFNLIGESS  IRCTSDPQGN  GVWSSPAPRC  ELSVPAACPH
1851    PPKIQNGHYI  GGHVSLYLPG  MTISYTCDPG  YLLVGKGFIF  CTDQGIWSQL
                    *
1901    DHYCKEVNCS  FPLFMNGISK  ELEMKKVYHY  GDYVTLKCED  GYTLEGSPWS
1951    QCQADDRWDP  PLAKCTSRAH  DALIVGTLSG  TIFFILLIIF  LSWIILKHRK
2001    GNNAHENPKE  VAIHLHSQGG  SSVHPRTLQT  NEENSRVLP
```

Domain structure
Consists of four long homologous repeats (LHRs) A,B,C,D, each containing a series of seven tandemly arranged short consensus repeats (SCRs) of 60–70 amino acids (each characterized by four conserved cysteines).

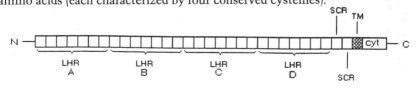

CD35

1–41	Signal sequence
42–489	LHR A
492–941	LHR B
942–1394	LHR C
1395–1846	LHR D
1972–1996	Transmembrane domain
1997–2039	Cytoplasmic domain

Twenty potential N-linked glycosylation sites.

Database accession numbers

	PIR	SWISSPROT	EMBL/GENBANK
Human	A28507	P17927	Y00816

Alternative forms

Four allotypes of CD35: A or F, B or S, C or F' and D, each of which can mediate binding to C3b [2] and which differ in the number of C3b binding sites. The efficiency with which erythrocytes bind immune complexes depends on the allele expressed. An increase in the copy number of CD35 and in the number of C3b binding sites increases the affinity for C3b while at the same time increasing the inhibitory effect on C3 convertases [7,12].

References
[1] Klickstein, L.B. et al. (1987) J. Exp. Med. 165, 1095–1112.
[2] **Ahearn, J.M. and Fearon, D.T. (1989) Adv. Immunol. 46, 183–219.**
[3] Iida, K. and Nussenzweig, V. (1981) J. Exp. Med. 153, 1138–1150.
[4] Neuman, E. et al. (1990) J. Immunol. 145, 3325–3332.
[5] Berger, M. et al. (1991) Proc. Natl Acad. Sci. USA 88, 3019–3023.
[6] Klickstein, L.B. et al. (1988) J. Exp. Med. 168, 1699–1717.
[7] Krych, M. et al. (1992) Curr. Opinion Immunol. 4, 8–13.
[8] Kalli, K. et al. (1991) J. Exp. Med. 174, 1451–1460.
[9] Wong, W. et al. (1989) J. Exp. Med. 169, 847–863.
[10] Weis, J.H. et al. (1987) J. Immunol. 138, 312
[11] Carroll, M.C. et al. (1988) J. Exp. Med. 167, 1271–1280.
[12] Wong, W. and Farrell, S.A. (1991) J. Immunol. 146, 656–662.

CD36 Platelet glycoprotein IV, GPIIIb, OKM5 antigen

Other names
Milk fat globule membrane (MFGM) PAS IV

Family
Orphan.

Cellular distribution
Small vessel endothelium, platelets, monocytes, megakaryocytes, reticulocytes, mammary epithelial cells [1].

Function
For a recent review see ref. 2.

CD36 directly mediates cytoadherence of *Plasmodium falciparum* parasitized erythrocytes to post-capillary vessels in lung, kidney, heart, placenta and brain [3]. It also mediates the divalent cation-dependent adherence of platelets to thrombospondin and collagen [4–6].

CD36 has been reported to be physically associated with *fyn*, *lyn* and *yes* tyrosine kinases [7].

Regulation of expression
Generally constitutively expressed although shown to increase in myeloproliferative disorders. Also upregulated on U937 cells in response to phorbol ester [4–6].

Ligands
Collagen [6]. Thrombospondin (through its CSVTCG sequence) [1,8]. Sequestrin [9].

Gene structure
At least 12 exons (R.H. Lipsky et al., unpublished data).

Gene location
Not known.

Molecular weights
Polypeptide 52 922
SDS Page 88 000, subject to tissue and species-specific glycosylation [10].

CD36

Amino acid sequence (from human placenta) [4]

```
  1  GCDRNCGLIA GAVIGAVLAV FGGILMPVGD LLIQKTIKKQ VVLEEGTIAF
                                *
 51  KNWVKTGTEV YRQFWIFDVQ NPQEVMMNSS NIQVKQRGPY TYRVRFLAKE
       *                            *
101  NVTQDAEDNT VSFLQPNGAI FEPSLSVGTE ADNFTVLNLA VAAASHIYQN
                  *
151  QFVQMILNSL INKSKSSMFQ VRTLRELLWG YRDPFLSLVP YPVTTTVGLF
        *          *                 *              *
201  YPYNNTADGV YKVFNGKDNI SKVAIIDTYK GKRNLSYWES HCDMINGTDA
251  ASFPPFVEKS QVLQFFSSDI CRSIYAVFES DVNLKGIPVY RFVLPSKAFA
                 *
301  SPVENPDNYC FCTEKIISKN CTSYGVLDIS KCKEGRPVYI SLPHFLYASP
351  DVSEPIDGLN PNEEEHRTYL DIEPITGFTL QFAKRLQVNL LVKPSEKIQV
              *
401  LKNLKRNYIV PILWLNETGT IGDEKANMFR SQVTGKINLL GLIEMILLSV
451  GVVMFVAFMI SYCACRSKTI K
```

Domain structure
Not defined
  1–437   Extracellular domain
  438–465 Transmembrane domain (based on cDNA sequence)
  466–471 Cytoplasmic domain

The N-terminal contains a hydrophobic signal sequence which is not cleaved. The initiator methionine found in the cDNA sequence is not present in the mature protein.

Ten potential N-linked glycosylation sites.

Database accession numbers

	PIR	SWISSPROT	EMBL/GENBANK	REFERENCE
Human	A30989	P16671	M24795	4

Alternative forms

MFGM PAS IV appears to be an identical polypeptide to CD36 which undergoes cell-type-specific glycosylation yielding an 80 kD form [10].

References

1. Asch, A.S. et al. (1987) J. Clin. Invest. 79, 1054–1061.
2. **Catimel, B. et al. (1991) In 'Platelet Immunology: Fundamental and Clinical Aspects' (Kaplan-Gouet, N. et al. eds), INSERM/John Libbey Eurotext Ltd, Paris, pp. 41–53.**
3. Barnwell, J.W. et al. (1989) J. Clin. Invest. 765–772.
4. **Oquendo, P. et al. (1989) Cell, 58, 95–101.**
5. Ockenhouse, C.F. et al. (1989) Science 243, 1469–1471.
6. Tandon, N.N. et al. (1989) J. Biol. Chem. 264, 7570–7575.
7. Huang, M-M. et al. (1991) Proc. Natl Acad. Sci. USA 88, 7844–7848.
8. Asch, A.S. et al. (1992) Biochem. Biophys. Res. Commun. 182, 1208–1217.
9. Ockenhouse, C.F. et al. (1991) Proc. Natl Acad. Sci. USA 88, 3175–3179.
10. Greenwalt, G.E. et al. (1990) Biochemistry 29, 7054–7059.

CD44

H-CAM, GP90[HERMES]**, Pgp-1, ECRM III**

Other names
In(Lu)-related p80, HUTCH-1, Ly-24, p85

Family
The N-terminal region of CD44 is related to the cartilage proteoglycan core and link proteins.

Cellular distribution
CD44 is broadly distributed. Amongst haematopoietic cells CD44 is expressed on B and T lymphocytes, monocytes and neutrophils. Other CD44-positive cell types include epithelial cells, glial cells, fibroblasts and myocytes [reviewed in ref.1].

Function
The broad distribution of CD44 suggests a general role in cell–cell and cell–matrix adhesion [reviewed in ref.1]. CD44 is involved in T cell activation and adhesion [2]. Triggering of CD44 increases homotypic T cell aggregation mediated via LFA-1 [3]. CD44 is linked to the cytoskeleton [4]. Although CD44 has been shown to be involved in lymphocyte adhesion to high endothelial venules (HEV) [5], it appears to function only in an accessory role.

Regulation of expression
Lymphocyte maturation is associated with changes in levels of CD44 [reviewed in ref.1]. Increased expression of CD44 is found on some carcinomas [6]. The transition of tumour cell lines from non-metastatic to metastatic, may be associated with the changes in the expression of CD44 variants [7]. A variant sharing sequence similarities with metastasis-promoting variants is transiently expressed on B and T lymphocytes following antigen stimulation [8].

Ligands
CD44H binds hyaluronate [9,10]. CD44 variants with attached chondroitin sulphate are able to bind fibronectin, laminin and collagen [11]. CD44 may also bind homotypically [12].

Gene location
11p13.

Structure (CD44H)

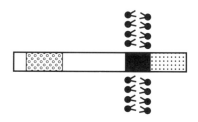

CD44

Molecular weight
Polypeptide 29 747
SDS PAGE 85 000–90 000 (CD44H), 110 000–160 000 (CD44E)

Amino acid sequence [13]

```
                         ↓       *
  1  MDKFWWHAAW GLCLVPLSLA QIDLNITCRF AGVFHVEKNG RYSISRTEAA
            *                                              *
 51  DLCKAFNSTL PTMAQMEKAL SIGFETCRYG FIEGHVVIPR IHPNSICAAN
               *          *
101  NTGVYILTSN TSQYDTYCFN ASAPPEEDCT SVTDLPNAFD GPITITIVNR

151  DGTRYVQKGE YRTNPEDIYP SNPTDDDVSS GSSSERSSTS GGYIFYTFST

201  VHPIPDEDSP WITDSTDRIP ATRDQDTFHP SGGSHTTHGS ESDGHSHGSQ
              *
251  EGGANTTSGP IRTPQIPEWL IILASLLALA LILAVCIAVN SRRS
```

Domain structure
 1–20 Signal sequence
 31–121 Link
 269–289 Transmembrane domain

Six potential N-linked glycosylation sites.

Alternatively, spliced variants are generated by the insertion of either a single exon, or a combination of exons, after residue 222.

Database accession numbers

	PIR	SWISSPROT	EMBL/GENBANK	REFERENCE
Human CD44H	A32377	P16070	M25078	
Human CD44E	S13530	P22511	X55150	10
Mouse CD44	A34424	P15379	M27129	14
Rat CD44			M61874	18
Rat CD44-pMeta-1			M61875	18

Alternative forms

Alternative splicing of at least eleven exons produces a number of variants [16] including CD44E (CD44R1), which is predominantly expressed on epithelial cells [15,17]. Post-translational modifications include variations in the pattern of N- and O-linked glycosylation and the addition of chondroitin sulphate to produce high molecular weight variants. These modifications account for the wide range of molecular weights observed, from M_r 85 000 to M_r 250 000.

A variant with an extended cytoplasmic tail has been reported, which is probably the predominant form [6]. In this form residue 294 is replaced by R and continues for an additional 67 residues.

CD44

```
 1  CGQKKKLVIN SGNGAVEDRK PSGLNGEASK SQEMVHLVNK ESSETPDQFM

51  TADETRNLQN VDMKIGV
```

CD44E (CD44R1) [10] contains an additional 132 amino acids inserted in the extracellular domain.

CD44-pMeta-1 [18] is reported to be expressed in some metastasizing, but not non-metastasizing tumour cell lines.

References

[1] **Gallatin, W.M. et al. (1991) In Cellular and Molecular Mechanisms of Inflammation (Cochrane, C.G. and Gimbrone, M.A. (eds) Academic Press, London.**
[2] Shimizu, Y. et al. (1989) J. Immunol. 143, 2457–2463.
[3] Koopman, G. et al. (1990) J. Immunol. 145, 3589–3593.
[4] Kalomiris, E.L. and Bourguignon, L.Y.W. (1988) J. Cell Biol. 106, 319–327.
[5] Jalkenen, S. et al. (1987) J. Cell Biol. 105, 983–990.
[6] Stamenkovic, I. et al. (1989) Cell 56, 1057–1062.
[7] Hoffman, M. et al. (1991) Cancer Res. 51, 5292–5297.
[8] Arch, R. et al. (1992) Science 257, 682–685.
[9] Aruffo, A. et al. (1990) Cell 61, 1303–1313.
[10] Stamenkovic, I. et al. (1991) EMBO J. 10, 343–348.
[11] Jalkenen, S. and Jalkenen, M. (1992) J. Cell Biol. 116, 817–825
[12] Belitsos, P.C. et al. (1990) J. Immunol. 144, 1661–1670.
[13] Goldstein, L.A. et al. (1989) Cell 56, 1063–1072.
[14] Zhou, D.F. et al. (1990) J. Immunol. 143, 3390–3395.
[15] Brown, T.A. et al. (1991) J. Cell Biol. 113, 207–221.
[16] Jackson, D.G. et al. (1992) J. Biol. Chem. 267, 4732–4739.
[17] Dougherty, G.J. et al. (1991) J. Exp. Med. 174, 1–5.
[18] Gunthert, U. et al. (1991) Cell 65, 13–24.

CEA carcinoembryonic antigen

Family
CEA is one of eight members of the CEA family, which is a subset of the immunoglobulin superfamily

Cellular distribution
CEA is expressed in tissues derived from all three germ layers during embryogenesis. CEA is present at low levels on adult colon epithelial cells, but greatly over-expressed on the majority of colon carcinomas. CEA is also present on a variety of other epithelial-derived tumours including breast and lung [reviewed in ref.1].

Function
Transfected cells expressing CEA undergo specific calcium-independent aggregation and homotypic cell sorting [2]. The role of CEA in intercellular adhesion *in vivo* is unproven.

Regulation of expression
The expression of CEA is developmentally regulated, being present during embryogenesis, but only detectable in small amounts in normal adult tissues. It is, however, strongly re-expressed on the majority of colon carcinomas and a high proportion of carcinomas from other tissues.

Ligand
CEA mediated binding in transfected cells is homophilic [2]. CEA binds weakly to NCA [3]. An M_r 80 000 ligand has been identified in Kuppfer cells of the liver [4].

Gene structure[5]

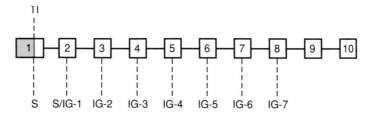

Exon 9 encodes the sequence for the attachment of GPI-anchor.

Gene location and size
19q 13.1–13.3; 24 kb.

Structure

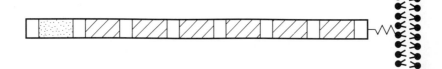

CEA

Molecular weights

Polypeptide   70 493
SDS PAGE   180 000

Amino acid sequence (from LS180 colon adenocarcinoma) [6]

```
              ↓
  1  MESPSAPPHR WCIPWQRLLL TASLLTFWNP PTTAKLTIES TPFNVAEGKE
 51  VLLLVHNLPQ HLFGYSWYKG ERVDGNRQII GYVIGTQQAT PGPAYSGREI
              *          *
101  IYPNASLLIQ NIIQNDTGFY TLHVIKSDLV NEEATGQFRV YPELPKPSIS
              *                             *           *
151  SNNSKPVEDK DAVAFTCEPE TQDATYLWWV NNQSLPVSPR LQLSNGNRTL
              *  *                                       *
201  TLFNVTRNDT ASYKCETQNP VSARRSDSVI LNVLYGPDAP TISPLNTSYR
              *                  *              *        *
251  SGENLNLSCH AASNPPAQYS WFVNGTFQQS TQELFIPNIT VNNSGSYTCQ
              *          *
301  AHNSDTGLNR TTVTTITVYA EPPKPFITSN NSNPVEDEDA VALTCEPEIQ
              *         *           *
351  NTTYLWWVNN QSLPVSPRLQ LSNDNRTLTL LSVTRNDVGP YECGIQNELS
                                                *
401  VDHSDPVILN VLYGPDDPTI SPSYTYYRPG VNLSLSCHAA SNPPAQYSWL
                         *           *
451  IDGNIQQHTQ ELFISNITEK NSGLYTCQAN NSASGHSRTT VKTITVSAEL
              *                     *
501  PKPSISSNNS KPVEDKDAVA FTCEPEAQNT TYLWWVNGQS LPVSPRLQLS
              *  *                  *
551  NGNRTLTLFN VTRNDARAYV CGIQNSVSAN RSDPVTLDVL YGPDTPIISP
                         *                                *
601  PDSSYLSGAN LNLSCHSASN PSPQYSWRIN GIPQQHTQVL FIAKITPNNN
                         *
651  GTYACFVSNL ATGRNNSIVK SITVSASGTS PGLSAGATVG IMIGVLVGVA
701  LI
```

Domain structure

  1–34   Signal sequence
 51–124   Ig-1
160–219   Ig-2
252–303   Ig-3
338–387   Ig-4
430–481   Ig-5
516–574   Ig-6
608–659   Ig-7

Residues 104, 115 and 152 are confirmed N-linked glycosylation sites. There are 25 additional potential sites. Ala 676 is the attachment site for the GPI-anchor.

Database accession numbers

	PIR	SWISSPROT	EMBL/GENBANK
Human	A36319	P06731	M15042

Alternative forms

Variants differ in the extent of glycosylation.

References
[1] Shuster, J.D. et al. (1980) Prog. Exp. Tumour Res. 15, 89–139.
[2] **Benchimol, S. et al. (1989) Cell 57, 327–334.**
[3] Oikawa, S. et al. (1989) Biochem. Biophys. Res. Commun. 164, 39–45.
[4] Toth, C.A. et al. (1985) Cancer Res. 45, 392–397.
[5] Schrewe, H. et al. (1990) Mol. Cell. Biol. 10, 2738–2748.
[6] Beauchemin, N. et al. (1987) Mol. Cell. Biol. 7, 3221–3230.

Cell-CAM 105   C-CAM

Family
Cell-CAM 105 is a member of the immunoglobulin superfamily and an ecto-ATPase [1].

Cellular distribution
Cell-CAM 105 is present on epithelial and endothelial cells, megakaryocytes, activated platelets, neutrophils and monocytes [2].

Function
Cell-CAM 105 has been shown to function as a homophilic, calcium-independent adhesion molecule [3]. Cell-CAM 105 binds calmodulin [4].

Regulation of expression
Cell-CAM 105 is generally expressed in the later developmental stages. Expression of Cell-CAM 105 is transiently downregulated following hepatectomy and is altered in hepatocellular carcinomas [reviewed in ref.5].

Ligand
Cell-CAM 105 mediated adhesion is homophilic [3].

Structure

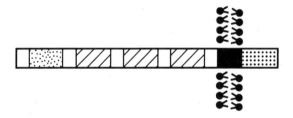

The N-terminal resembles an Ig V-domain but lacks the characteristic disulphide bonds.

Molecular weights
Polypeptide   53 627
SDS PAGE   105 000

Cell-CAM 105

Amino acid sequence [6]

```
                                           ↓
  1   MELASARLLR GQIPWRGLLL TASLLTYWSP LTTAQVTVDA VPPNVVEEKS
                                            *
 51   VLLLAHNLPQ EFQVFYWYKG TTLNPDSEIA RYIRSDNMSK TGPAYSGRET
           *          *                                  *
101   IYSNGSLFFQ NVNKTDERAY TLSVFDQQFN PIQTSVQFRV YPALQKPNVT
                                ** *                     *
151   GNNSNPMEGE PFVSLMCEPY TNNTSYLWSR NGESLSEGDR VTFSEGNRTL
                                 *
201   TLLNVRRTDK GYYECEARNP ATFNRSDPFN LDVIYGPDAP VISPPDIYLH
           *                                *         *
251   QGSNLNLSCH ADSNPPAQYF WLINEKLQTS SQELFISNIT TNNSGTYACF
           *          *          *
301   VNNTVTGLSR TTVKNITVFE PVTQPSIQIT NTTVKELGSV TLTCFSKDTG
                                *
351   VSVRWLFNSQ SLQLTDRMTL SQDNSTLRID PIKREDAGDY QCEISNPVSF

401   RISHPIKLDV IPDPTQGNSG LSEGAIAGIV IGSVAGVALI AALAYFLYSR

451   KTGGGSDHRD LTEHKPSTSS HNLGPSDDSP NKVDDVSYSV LNFNAQQSKR

501   PTSASSSPTE TVYSVVKKK
```

Domain structure
17–34   Signal sequence
42–140 Ig-1
159–219 Ig-2
252–303 Ig-3
338–396 Ig-4
423–448 Transmembrane domain

Sixteen potential N-linked glycosylation sites.

Database accession numbers

	PIR	SWISSPROT	EMBL/GENBANK
Rat		P16573	J04963

References
[1] Aurivillius, M. et al. (1990) FEBS. Lett. 264, 267–269.
[2] Odin, P. and Obrink, B. (1987) Exp. Cell Res. 171, 1–15.
[3] Tingstrom, A. et al. (1990) J. Cell Sci. 96, 17–25.
[4] Blikstad, I. et al. (1992) FEBS Lett. 302, 26–30.
[5] **Obrink, B. (1991) Bioessays 13, 227–234.**
[6] Lin, S-H. and Guidotti, G. (1989) J. Biol. Chem. 264, 14408–14414.

Contactin   F11 [1]

Family
Immunoglobulin superfamily, C2 subset. The extracellular region of contactin shows strong homology with mouse F3.

Cellular distribution
Contactin is expressed on neurons and is localized predominantly to axons.

Function
The structural similarities of contactin with known cell adhesion molecules, together with its restricted expression, suggest a role in neuronal interactions and neurite fasciculation.

Regulation of expression
The expression of contactin is developmentally regulated. It is maximally expressed in mature tissue.

Ligand
Contactin binds tenascin [2].

Structure

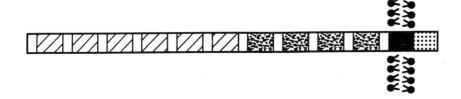

Molecular weights

Polypeptide	119 956
SDS PAGE	130 000

Amino acid sequence (from 13 day chick embryo) [3]

```
                              ↓
  1  MRFFISHLVT  LCFIFCVADS  THFSEEGNKG  YGPVFEEQPI  DTIYPEESSD

 51  GQVSMNCRAR  AVPFPTYKWK  LNNWDIDLTK  DRYSMVGGRL  VISNPEKSRD

101  AGKYVCVVSN  IFGTVRSSEA  TLSFGYLDPF  PPEEHYEVKV  REGVGAVLLC
                                                             *
151  EPPYHYPDDL  SYRWLLNEFP  VFIALDRRRF  VSQTNGNLYI  ANVEASDKGN
                                                             *
201  YSCFVSSPSI  TKSVFSKFIP  LIPQADRAKV  YPADIKVKFK  DTYALLGQNV
```

Contactin

```
 251   TLECFALGNP VPELRWSKYL EPMPATAEIS MSGAVLKIFN IQYEDEGLYE
                                     *
 301   CEAENYKGKD KHQARVYVQA SPEWVEHIND TEKDIGSDLY WPCVATGKPI
 351   PTIRWLKNGV SFRKGELRIQ GLTFEDAGMY QCIAENAHGI IYANAELKIV
                                                           *
 401   ASPPTFELNP MKKKILAAKG GRVIIECKPK AAPKPKFSWS KGTELLVNGS
                  *                    *
 451   RIHIWDDGSL EIINVTKLDE GRYTCFAENN RGKANSTGVL EMTEATRITL
                  *
 501   APLNVDVTVG ENATMQCIAS HDPTLDLTFI WSLNGFVIDF EKEHEHYERN
                                       *
 551   VMIKSNGELL IKNVQLRHAG RYTCTAQTIV DNSSASADLV VRGPPGPPGG
 601   IRIEEIRDTA VALTWSRGTD NHSPISKYTI QSKTFLSEEW KDAKTEPSDI
 651   EGNMESARVI DLIPWMEYEF RIIATNTLGT GEPSMPSQRI RTEGAPPNVA
 701   PSDVGGGGGS NRELTITWMP LSREYHYGNN FGYIVAFKPF GEKEWRRVTV
 751   TNPEIGRYVH KDESMPPSTQ YQVKVKAFNS KGDGPFSLTA VIYSAQDAPT
 801   EVPTDVSVKV LSSSEISVSW HHVTEKSVEG YQIRYWAAHD KEAAAQRVQV
 851   SNQEYSTKLE NLKPNTRYHI DVSAFNSAGY GPPSRTIDII TRKAPPSQRP
                                     *
 901   RIISSVRSGS RYIITWDHVK AMSNESAVEG YKVLYRPDGQ HEGKLFSTGK
 951   HTIEVPVPSD GEYVVEVRAH NEGGDGEVAQ IKISGATAGV PTLLLGLVLP
1001   APRCPCPTRD SECSRVYLVL TRTQGTPFEE EDCNSRFLWH HPKPNKLHFK
1051   ISYPTNLLWT LKGLRHPGTP SLSKSTFERI RNIFNLFQYA L
```

Domain structure
1–20 Signal sequence
51–110 Ig-1
143–207 Ig-2
247–305 Ig-3
336–386 Ig-4
420–479 Ig-5
510–578 Ig-6
615–701 FN-III
718–799 FN-III
820–896 FN-III
916–976 FN-III
983–1002 Transmembrane domain

Nine potential N-linked glycosylation sites.

Contactin

Database accession numbers
	PIR	SWISSPROT	EMBL/GENBANK
Chicken	S01998	P10450	Y00813

References
[1] Brummendorf, P. (1989) Neuron 2, 1351–1361.
[2] Zisch, A.H. et al. (1992) J. Cell Biol. 119, 203–213.
[3] Ranscht, B. (1989) J. Cell Biol. 107, 1561–1573.

E-Cadherin

Uvomorulin (mouse), ARC-1 (dog)

Other names
Cell CAM120/180 (human and mouse), L-CAM or liver cell adhesion molecule (chicken).

Family
Cadherin.

Cellular distribution
Non-neural epithelial tissue.

Function
Ca^{2+} dependent adhesion between epithelial cells [1]. E-Cadherin is involved in the generation and maintenance of epithelial layers during development and in adult tissues [2]. It is also proposed to be involved in the compaction of pre-implantation embryos in mouse [3,4]. It may also have a role in the induction of cell surface polarity of certain cytoplasmic and membrane proteins [5]. Clustering of E-Cadherin is thought to occur via α, β and γ catenins which link E-Cadherin to cytoskeletal actin [6–8]. The E-Cadherin (uvomorulin)–catenin complex formation is crucial for regulating cell adhesiveness and also for linking E-Cadherin to other integral membrane proteins.

Regulation of expression
Developmentally regulated. May be associated with the transformation state or metastatic potential of certain cells [1].

Ligand
E-Cadherin (homotypic adhesion) via an HAV sequence located in the N-terminal 113 amino acids.

Gene structure

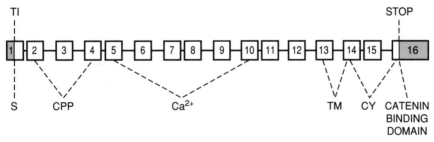

exon
1   5'UT and signal peptide
2–4 Cleaved protein precursor
4–13 Extracellular domain
5–10 Contain segments of internal homology with calcium binding motifs
13   Transmembrane domain
14   Transmembrane and cytoplasmic domain
15   Cytoplasmic domain
16   Cytoplasmic domain and 3'UT

E-Cadherin

Mouse gene (shown below) contains 16 exons spanning over 40 kb [9] which is nearly identical to the chicken L-CAM gene which spans 9 kb and also contains 16 exons (see ref. 10 and also entry for B-Cadherin).

Gene location and size
8 (mouse); 40 kb [11]
16 p11-16qter (human) [12].

Molecular weights
Polypeptide   80 317 (mouse)
SDS-PAGE: 135 000 (precursor); 124 000 (on cell surface)

Amino acid sequence (from mouse teratocarcinoma F9 cells) [1]

```
  1  MGARCRSFSA LLLLLQVSSW LCQELEPESC SPGFSSEVYT FPVPERHLER
 51  GHVLGRVRFE GCTGRPRTAF FSEDSRFKVA TDGTITVKRH LKLHKLETSF
101  LVRARDSSHR ELSTKVTLKS MGHHHHRHHH RDPASESNPE LLMFPSVYPG
151  LRRQKRDWVI PPISCPENEK GEFPKNLVQI KSNRDKETKV FYSITGQGAD
201  KPPVGVFIIE RETGWLKVTQ PLDREAIAKY ILYSHAVSSN GEAVEDPMEI
251  VITVTDQNDN RPEFTQEVFE GSVAEGAVPG TSVMKVSATD ADDVNTYNA
301  AIAYTIVSQD PELPHKNMFT VNRDTGVISV LTSGLDRESY PTYTLVVQAA
351  DLQGEGLSTT AKAVITVKDI NDNAPVFNPS TYQGQVPENE VNARIATLKV
401  TDDDAPNTPA WKAVYTVVND PDQQFVVVTD PTTNDGILKT AKGLDFEAKQ
451  QYILHVRVEN EEPFEGSLVP STATVTVDVV DVNEAPIFMP AERRVEVPED
501  FGVGQEITSY TAREPDTFMD QKITYRIWRD TANWLEINPE TGAIFTRAEM
551  DREDAEHVKN STYVALIIAT DDGSPIATGT GTLLLVLLDV NDNAPIPEPR
601  NMQFCQRNPQ PHIITILDPD LPPNTSPFTA ELTHGASVNW TIEYNDAAQE
651  SLILQPRKDL EIGEYKIHLK LADNQNKDQV TTLDVHVCDC EGTVNNCMKA
701  GIVAAGLQVP AILGILGGIL ALLILILLLL LFLRRRTVVK EPLLPPDDDT
751  RDNVYYYDEE GGGEEDQDFD LSQLHRGLDA RPEVTRNDVA PTLMSVPQYR
801  PRPANPDEIG NFIDENLKAA DSDPTAPPYD SLLVFDYEGS GSEAASLSSL
851  NSSESDQDQD YDYLNEWGNR FKKLADMYGG GEDD
```

Domain structure

1–27	Signal sequence
28–156	Propeptide
157–709	Extracellular domain
167–264	Repeat I
265–377	Repeat II
378–488	Repeat III
489–595	Repeat IV
596–699	Repeat V
710–733	Transmembrane domain
734–884	Cytoplasmic domain
153–156	Protease cleavage recognition site [13]

Five potential N-linked glycosylation sites (four in extracellular domain only shown).

Three Ca^{2+} binding sites in the extracellular domain (residues 256–260, 290–292 and 369–372).

The C-terminal 72 amino acids are necessary for the interaction with catenins [14].

Database accession numbers

	PIR	SWISSPROT	EMBL/GENBANK	REFERENCE
Human	S05475			
Human, partial sequence		P12830		[12]
Mouse	S04528		X12790 X06115 X06339	
Chicken		P09803 P08641		[1] [15]
Chicken			M16260	

References

[1] **Nagafuchi, A. et al. (1987) Nature 329, 341–343.**
[2] Kemler, R. et al. (1989) Curr. Opinion Cell Biol. 1, 892–897.
[3] Kemler, R. et al. (1977) Proc. Natl Acad. Sci. USA 74, 4449–4452.
[4] Hyafil, F.C. et al. (1981) Cell 26, 447–454.
[5] McNeill, H. et al. (1990) Cell 62, 309–316.
[6] **Ranscht, B. (1991) in Seminars in the Neurosciences 3, 285–296.**
[7] Ozawa, et al. (1989) EMBO J. 8, 1711–1717.
[8] Ozawa, M. and Kemler, R. (1992) J. Cell Biol. 116, 989–996.
[9] Ringwald, M. et al. (1991) Nucleic Acids Res. 19, 6533–6539.
[10] Sorkin, B.C. et al. (1988) Proc. Natl Acad. Sci. USA 85, 7617–7621.
[11] Eistetter, H. et al. (1988) Proc. Natl Acad. Sci. USA 85, 3489–3493.
[12] Mansouri, A. et al. (1988) Differentiation 38, 67–71.
[13] Ozawa, M. and Kemler, R. (1990) J. Cell Biol. 111, 1645–1650.
[14] Nagafuchi, A. and Takeichi, M. (1989) Cell Regul. 1, 37–44.
[15] Gallin, W. et al. (1987) Proc. Natl Acad. Sci. USA 84, 2808–2812.

E-Selectin

ELAM-1 (endothelial leukocyte adhesion molecule 1) [1], LECAM-2

Family
Selectin (Ca^{2+} dependent, C-type lectin).

Cellular Distribution
Endothelium.

Function
E-Selectin mediates neutrophil [2–4], monocyte [5,6] and some memory T cell (CD4+) [7,8] adhesion to vascular endothelium hence is important in the regulation of inflammatory and immunological events at the vessel wall. Initial binding of leukocytes by E-selectin is thought to trigger the recruitment and activation of additional adhesion molecules to the site of inflammation, such as Mac-1, by as yet unknown mechanisms [9].

E-Selectin may also play a role in the metastasis of tumour cells since it has been shown to mediate binding of the human carcinoma cell line HT-29 to cytokine activated endothelium [10] and a number of tumour cell lines, bearing carbohydrates which bind to E-selectin, have also been identified [11,12].

Regulation of expression
E-Selectin is expressed at sites of inflammation *in vivo* [13]. Immunohistochemical studies of human tissue in a variety of disease states show E-selectin expression on vascular endothelium, particularly in inflammatory diseases showing substantial neutrophil infiltration [14–16]. Activation of endothelium by cytokines such as IL1β and TNFα and LPS results in rapid though transient upregulation of E-selectin reaching a peak at 4–6 h and declining to baseline levels by 24–48 h *in vitro* [2,17].

Ligand
Sialyl Lewis^x, sialyl Lewis^a and related fucosylated N-acetyl-lactosamines found on leukocyte glycolipids and glycoproteins may act as ligands for ELAM-1 [11,12,18–23]. Recent studies have also shown that *in vitro*, the density of expression of E-selectin at the cell surface has a marked influence on carbohydrate binding [24].

Cutaneous lymphocyte antigen (CLA) has been identified as a carbohydrate-bearing homing receptor for E-selectin expressed at cutaneous sites of chronic inflammation [25].

Gene structure

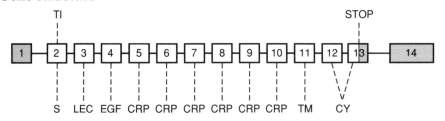

exon		exon	
1	Promotor + 5'UT	5	CRP1
2	Signal peptide	6	CRP2
3	Lectin domain	7	CRP3
4	EGF domain	8	CRP4

E-Selectin

9   CRP5
10 CRP6
11 Transmembrane
12 Cytoplasmic domain 1
13 Cytoplasmic domain 2 and 3'UT
14 3'UT

Sp\ans 13 kb with 13 introns and 14 exons. Distinct exons encode each protein structural domain [26]. The promotor contains NFkB and AP1 binding sites [27]. An additional sequence for the putative NF-ELAM binding protein has also been identified [28].

Gene location and size
1q12-qter; 13 kb [26]

Structure

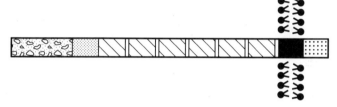

Molecular weights
Polypeptide   64 466
SDS-PAGE   107 000 and 115 000 (different glycosylated forms).

Amino acid sequence (from IL1β stimulated HUVECs [2]).

```
              ↓   *
  1   MIASQFLSAL TLVLLIKESG AWSYNTSTEA MTYDEASAYC QQRYTHLVAI

 51   QNKEEIEYLN SILSYSPSYY WIGIRKVNNV WVWVGTQKPL TEEAKNWAPG
                                                          *
101   EPNNRQKDED CVEIYIKREK DVGMWNDERC SKKKLALCYT AACTNTSCSG
                *                    *                    *
151   HGECVETINN YTCKCDPGFS GLKCEQIVNC TALESPEHGS LVCSHPLGNF
         *
201   SYNSSCSISC DRGYLPSSME TMQCMSSGEW SAPIPACNVV ECDAVTNPAN
                           *
251   GFVECFQNPG SFPWNTTCTF DCEEGFELMG AQSLQCTSSG NWDNEKPTCK
                           *                    *
301   AVTCRAVRQP QNGSVRCSHS PAGEFTFKSS CNFTCEEGFM LQGPAQVECT

351   TQGQWTQQIP VCEAFQCTAL SNPERGYMNC LPSASGSFRY GSSCEFSCEQ

401   GFVLKGSKRL QCGPTGEWDN EKPTCEAVRC DAVHQPPKGL VRCAHSPIGE

451   FTYKSSCAFS CEEGFELHGS TQLECTSQGQ WTEEVPSCQV VKCSSLAVPG
         *                         *
501   KINMSCSGEP VFGTVCKFAC PEGWTLNGSA ARTCGATGHW SGLLPTCEAP

551   TESNIPLVAG LSAAGLSLLT LAPFLLWLRK CLRKAKKFVP ASSCQSLESD

601   GSYQKPSYIL
```

E-Selectin

Domain structure
1–21	Signal sequence
22–141	Lectin domain
142–176	EGF domain
177–238	CRP repeat
239–300	CRP repeat
301–363	CRP repeat
364–426	CRP repeat
427–489	CRP repeat
490–548	CRP repeat
557–578	Transmembrane domain
579–610	Cytoplasmic domain

Rabbit ELAM-1 contains only five CRP repeats [29].

Eleven potential N-linked glycosylation sites.

The cytoplasmic domain contains eight potential phosphorylation sites (6 Ser and 2 Tyr).

Database accession numbers
	PIR	SWISSPROT	EMBL/GENBANK	REFERENCE
Human	A32606	P16581	M30640	
Mouse		P16111	M80777	[30]
			M80778	[30]

Alternative forms
Unknown.

References
[1] Bevilacqua, M.P. et al. (1991) Cell 67, 233.
[2] **Bevilacqua, M.P. et al. (1989) Science 243, 1160–1165.**
[3] Hakkert, B.C. et al. (1991) Blood 78, 2721–2726.
[4] Hession, C.L. et al. (1990) Proc. Natl Acad. Sci. USA 87, 1673–1677.
[5] Leeuwenberg, J.F.M. et al. (1992) Scand. J. Immunol. 35, 335–341.
[6] Carlos, T.N. et al. (1991) Blood 77, 2266–2726.
[7] Picker, L.J. et al. (1991) Nature 349, 796–799.
[8] Shimizu, Y. et al. (1991) Nature 349, 799–802.
[9] Lo, S.K. et al. (1991) J. Exp. Med. 173, 1493–1500.
[10] Aruffo, A. et al. (1992) Proc. Natl Acad. Sci. USA 82, 2292–2296.
[11] Walz, G. et al. (1990) Science 250, 1132–1135.
[12] Takada, A. et al. (1991) Biochem. Biophys. Res. Commun. 179, 713–719.
[13] Pober, J.S. and Cotran, R.S. (1991) Lab. Invest. 64, 301–305.
[14] Cotran, R.S. et al. (1986) J. Exp. Med. 164, 661–666.
[15] Koch, A.E. et al. (1991) Lab. Invest. 64, 313–320.
[16] Groves, R.W. et al. (1991) Br. J. Dermatol. 124, 117–123.
[17] Bevilacqua, M.P. et al. (1987) Proc. Natl Acad. Sci. USA 84, 9238–9242.
[18] Tiermeyer, M. et al. (1991) Proc. Natl Acad. Sci. USA 88, 1138–1142.
[19] Tyrrell, D. et al. (1991) Proc. Natl Acad. Sci. USA 88, 10372–10376.
[20] Lowe, J.B. et al. (1990) Cell 63, 475–484.

21 Phillips, M.L. et al. (1990) Science 250, 1130–1132.
22 Goelz, S.E. et al. (1990) Cell 63, 1349–1356.
23 Berg, E.L. et al. (1991) J. Biol. Chem. 266, 14869–14872.
24 Larkin, M. et al. (1992) J. Biol. Chem. 267, 13661–13668.
25 Berg, E.L. et al. (1991) J. Exp. Med. 174, 1461–1466.
26 Collins, T. (1991) J. Biol. Chem. 266, 2466–2473.
27 Montgomery, K.F. et al. (1991) Proc. Natl Acad. Sci. USA 88, 6523–6527.
28 Whelan, J. et al. (1991) Nucleic Acids Res. 19, 2645–2653.
29 Larigan, J.D. et al. (1992) DNA Cell Biol. 11, 149–162.
30 Becker, M. et al. (1992) Eur. J. Biochem. 206, 401–411.

F3

Family
Immunoglobulin superfamily, C2 subset. F3 shows strong homology with chick F11.

Cellular distribution
F3 is expressed on subpopulations of neurons and is localized predominantly to axons [1].

Function
The structural similarities of F3 with known cell adhesion molecules, together with its restricted expression, suggest a role in neuronal interactions and neurite fasciculation. F3 transfected cells promote neurite outgrowth over their surface [2] and a soluble form of F3 enhanced neurite initiation and outgrowth [3]. In suspension, transfected cells aggregate faster and form larger aggregates than controls [2].

Regulation of expression
Expression of F3 is developmentally regulated. Maximal expression occurs 1–2 weeks postnatally, and declines rapidly thereafter, although it is still present in adult tissues [2].

Ligand
Unknown.

Gene Location
15, band f.

Structure [2]

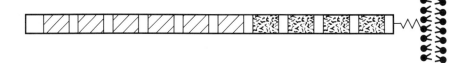

Molecular weights
Polypeptide   111 111
SDS PAGE   135 000

Amino acid sequence (from C57BL/6, postnatal day 2) [3]

```
   1  MKMPLLVSHL LLISLTSCLG DFTWHRRYGH GVSEEDKGFG PIFEEQPINT
  51  IYPEESLEGK VSLNCRARAS PFPVYKWRMN NGDVDLTNDR YSMVGGNLVI
 101  NNPDKQKDAG VYYCLASNNY GMVRSTEATL SFGYLDPFPP EERPRVKVKE
 151  GKGMVLLCDP PYHFPDDLSY RWLLNEFPVF ITMDKRRFVS QTNGNLYIAN
 201  VESSDRGNYS CFVSSPSITK SVFSKFIPLI PIPERTTKPY PADIVVQFKD
 251  IYTMMGQNVT LECFALGNPV PDIRWRKVLE PMPSTAEIST SGAVLKIFNI
 301  QLEDEGLYEC EAENIRGKDK HQARIYVQAF PEWVEHINDT EVDIGSDLYW
 351  PCIATGKPIP TIRWLKNGYS YHKGELRLYD VTFENAGMYO CIAENAYGSI
 401  YANAELKILA LAPTFEMNPM KKKILAAKGG RVIIECKPKA APKPKFSWSK
 451  GTEWLVNSSR ILIWEDGSLE INNITRNDGG IYTCFAENNR GKANSTGTLV
 501  ITNPTRIILA PINADITVGE NATMQCAASF DPALDLTFVW SFNGYVIDFN
 551  KEITHIHYQR NFMLSANGEL LIRNAQLKHA GRYTCTAQTI VDNSSASADL
 601  VVRGPPGPPG GLRIEDIRAT SVALTWSRGS DNHSPISKYT IQTKTILSDD
 651  WKDAKTDPPI IEGNMESAKA VDLIPWMEYE FRVVATNTLG TGEPSIPSNR
 701  IKTDGAAPNV APSDVGGGGG TNRELTITWA PLSREYHYGN NFGYIVAFKP
 751  FDGEEWKKVT VTNPDTGRYV HKDETMTPST AFQVKVKAFN NKGDGPYSLV
 801  AVINSAQDAP SEAPTEVGVK VLSSSEISVH WKHVLEKIVE SYQIRYWAGH
 851  DKEAAAHRVQ VTSQEYSARL ENLLPDTQYF IEVGACNSAG CGPSSDVIET
 901  FTRKAPPSQP PRIISSVRSG SRYIITWDHV VALSNESTVT GYKILYRPDG
 951  QHDGKLFSTH KHSIEVPIPR DGEYVVEVRA HSDGGDGVVS QVKISGVSTL
1001  SSSLLSLLLP SLGFLVYSEF
```

F3

Domain structure
1–20	Signal sequence
59–118	Ig-1
143–207	Ig-2
257–314	Ig-3
346–395	Ig-4
429–488	Ig-5
519–578	Ig-6
625–708	FN-1
728–810	FN-2
831–907	FN-3
927–986	FN-4
1004–1020	Transmembrane domain

Nine potential N-linked glycosylation sites.

Database accession numbers

	PIR	SWISSPROT	EMBL/GENBANK
Mouse	S05944	P12960	X14943

References
1. Faivre-Sarrailh, C. et al. (1992) J. Neurosci. 12, 257–267.
2. Gennarini, G. et al. (1991) Neuron 6, 595–606.
3. Durbec, P. et al. (1992) J. Cell Biol. 117, 877–889.
4. Gennarini, G. et al. (1989) J. Cell Biol. 109, 775–788.

Fasciclin I   fas I

Family
None known.

Cellular distribution
Fasciclin I is found on all PNS neurons, a subset of axon bundles of CNS neurons and on some non-neuronal cells during grasshopper and *Drosophila* development [1,2].

Function
Fasciclin I is involved in growth cone guidance and neurite fasciculation [3].

Regulation of expression
Expression of fasciclin I is developmentally regulated [7]. In addition, the amount of soluble fasciclin I increases during development, relative to the GPI-anchored form [5].

Ligand
Fasciclin I expressed in S2 cells functions as a homophilic adhesion molecule [4].

Gene location and size
89D; 14 kb.

Structure
The extracellular region of fasciclin I is divided into four repeat structures of approximately 150 amino acids which are attached to the membrane via a GPI-linkage [1,5].

Molecular weights
Polypeptide   70 674
SDS PAGE   70 000–72 000

Amino acid sequence [1]

```
  1   MLNAAALLLA LLCAANAAAA ADLADKLRDD SELSQFYSLL ESNQIANSTL
 51   SLRSCTIFVP TNEAFQRYKS KTAHVLYHIT TEAYTQKRLP NTVSSDMAGN
101   PPLYITKNSN GDIFVNNARI IPSLSVETNS DGKRQIMHII DEVLEPLTVK
151   AGHSDTPNNP NALKFLKNAE EFNVDNIGVR TYRSQVTMAK KESVYDAAGQ
201   HTFLVPVDEG FKLSARSSLV DGKVIDGHVI PNTVIFTAAA QHDDPKASAA
251   FEDLLKVTVS FFKQKNGKMY VKSNTIVGDA KHRVGVVLAE IVKANIPVSN
301   GVVHLIHRPL MIIDTTVTQF LQSFKENAEN GALRKFYEVI MDNGGAVLDD
```

Fasciclin I

```
                     *
351   INSLTEVTIL APSNEAWNSS NINNVLRDRN KMRQILNMHI IKDRLNVDKI
                *                                 *
401   RQKNANLIAQ VPTVNNNTFL YFNVRGEGSD TVITVEGGGV NATVIQADVA
                                                  *
451   QTNGYVHIID HVLGVPYTTV LGKLESDPMM SDTYKMGKFS HFNDQLNNTQ

501   RRFTYFVPRD KGWQKTELDY PSAHKKLFMA DFSYHSKSIL ERHLAISDKE

551   YTMKDLVKFS QESGSVILPT FRDSLSIRVE EEAGRYVIIW NYKKINVYRP

601   DVECTNGIIH VIDYPLLEEK DVVVAGGSYL PESSICIILA NLIMITVAKF

651   LN
```

Domain structure
1–21	Signal sequence
30–157	Domain 1
186–323	Domain 2
330–476	Domain 3
477–629	Domain 4

Five potential N-linked glycosylation sites.

Database accession numbers
	PIR	SWISSPROT	EMBL/GENBANK
Drosophila	B29900	P10674	M20545

Alternative forms

Three variants are produced by alternative splicing of two "micro-exons" [6].

References
[1] Zinn, K. et al. (1988) Cell 53, 577–587.
[2] **Greningloh, G. et al. (1990) Cold Spring Harbor Sym. Quant. Biol. 55, 327–340.**
[3] Jay, D.G. and Keshishian, H. (1990) Nature 348, 548–550.
[4] Elkins, T. et al. (1990) J. Cell Biol. 110, 1825–1832.
[5] Hortsch, M. and Goodman, C.S. (1990) J. Biol. Chem. 265, 15104–15109.
[6] McAllister, L. et al. (1992) J. Neurosci. 12, 895–905.
[7] McAllister, L. et al. (1992) Development 115, 267–276.

Fasciclin II   fas II

Family
Immunoglobulin superfamily, C2 subset. Fasciclin II is related to NCAM [for reviews see refs 1–3].

Cellular distribution
Fasciclin II is restricted to a subset of axons in the developing *Drosophila* and grasshopper embryos. There is limited expression outside the nervous system.

Function
Fasciclin II is involved in pathway recognition for axons during the development of nerve fascicles. In fasciclin II mutants, fascicles normally expressing the molecule fail to develop as a consequence of a failure in growth cone recognition [4].

Ligand
Fasciclin II functions as a homophilic adhesion molecule in transfected S2 cells [1].

Gene location
4B1-2.

Structure (transmembrane variant)

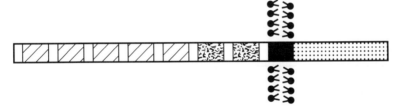

Molecular weights
Polypeptide   89 966 (GPI-linked variant); 96 816 (transmembrane variant)

Amino acid sequence [4]

```
  1   MGELPPNSVG VFLALLLCSC SLIELTRAQS PILEIYPKQE VQRKPVGKPL
                                  *
 51   ILTCRPTVPE PSLVADLQWK DNRNNTILPK PNGRNQPPMY TETLPGESLA

101   LMITSLSVEM GGKYYCTASY ANTEILEKGV TIKTYVAITW TNAPENQYPT

151   LGQDYVVMCE VKADPNPTID WLRNGDPIRT TNDKYVVQTN GLLIRNVQES
                                                                    *
201   DEGIYTCRAA VIETGELLER TIRVEVFIQP EIISLPTNLE AVEGKPFAAN

251   CTARGKPVPE ISWIRDATQL NVATADRFQV NPQTGLVTIS SVSQDDYGTY
                                  *
301   TCLAKNRAGV VDQKTKLNVL VRPQIYELYN VTGARTKEIA ITCRAKGRPA

351   PAITFRRWGT QEEYTNGQQD DDPRIILEPN FDEERGESTG TLRISNAERS
```

Fasciclin II

```
401  DDGLYQCIAR  NKGADAYKTG  HITVEFAPDF  SHMKELPPVF  SWEQRKANLS*
451  CLAMGIPNAT* IEWHWNGRKI  KDLYDTNLKI  VGTGPRSDLI  VHPVTRQYYS
501  GYKCIATNIH  GTAEHDMQLK  EARVPDFVSE  AKPSQLTATT  MTFDIRGPST
551  ELGLPILAYS  VQYKEALNPD  WSTAYNRSWS  PDSPYIVEGL  RPQTEYSFRF
601  AARNQVGLGN  WGVNQQQSTP  RRSAPEEPKP  LHNPVQHDKE  EPVVVSPYSD
651  HFELRWGVPA  DNGEPIDRYQ  IKYCPGVKIS  GTWTELENSC  NTVEVMETTS
701  FEMTQLVGNT  YYRIELKAHN  AIGYSSPASI  IMKTTRGESD  SANNNLGTLL
751  YSAGFNSGVG  ALHKRLFTTT  TTTTATSTTT  ITSITTATTT  IITLATTISI.
801  TLLSVLASML  A
```

Domain structure
1–28	Signal sequence
47–120	Ig-1
152–211	Ig-2
244–306	Ig-3
336–411	Ig-4
444–508	Ig-5
543–607	FN-III
656–723	FN-III

Database accession numbers

	PIR	SWISSPROT	EMBL/GENBANK
Drosophila GPI-linked variant			M77165
Drosophila transmembrane variant			M77166

Alternative forms

Membrane spanning and glycosylphosphotidylinositol (GPI)-linked forms are generated by alternative splicing[1,4].

The transmembrane variant has identical extracellular domains but diverges after residue 737:

```
  1  IDVIQVAERQ  VFSSAAIVGI  AIGGVLLLLF  VVDLLCCITV  HMGVMATMCR
 51  KAKRSPSEID  DEAKLGSGQL  VKEPPPSPLP  LPPPVKLGGS  PMSTPLDEKE
101  PLRTPTGSIK  QNSTIEFDGR  FVHSRSGEII  GKNSAV
```

15–32	Transmembrane domain

References

1 **Greningloh, G. et al. (1990) Cold Spring Harbor Symp. Quant. Biol. 55, 327–340.**
2 **Hortsch, M. and Goodman, C.S. (1991) Annu. Rev. Cell Biol. 505–557.**
3 Grenningloh, G. and Goodman, C.S. (1992) Curr. Opinions Cell Biol. 2, 42–47.
4 Grenningloh, G. et al. (1991) Cell 67, 45–57.

Fasciclin III — fas III

Family
Immunoglobulin superfamily.

Cellular distribution
Fasciclin III is expressed on a subset of axons and on some non-neuronal cells in the developing *Drosophila* CNS. There is some expression outside the nervous system.

Function
Fasciclin III is thought to be important in growth cone guidance [reviewed in refs 1,2].

Ligand
Fasciclin III functions as a homophilic adhesion molecule when expressed in S2 cells [3].

Gene location
36E1.

Structure [1]

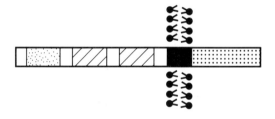

The N-terminal domain is V-like. The structure of the second and third domains is less clear, but they may resemble C2-like domains [1].

Molecular weights
Polypeptide   53 692
SDS PAGE   80 000 and 66 000

Fasciclin III

Amino acid sequence [3]

```
                          ↓
  1  MSRIVFICLA AILTDALTWA QVNVEPNTAL LNEGDRTELL CRYGRSINYC
                *
 51  RIEIPGEQKV LNLSPEWSKT PGFTYFGAGL TAGQCGVSIE RVKASNNGQV

101  KCSLGVEGEE LSGTIDLVVA LRPQQPIIEL LSRPNREGYF NEGTEFRARC
                *
151  SVRDGRPPAN ISWYIDNMPA NKRTTPLEVM SSTNDNVELS TSVQEIQWHL

201  SPEDSNRKLV CRSHHQTDRE SVPPQEAAYI INVRYAPVHQ PDAAVYGLYL
          *                                              *
251  EHTAIVNITI RASPQPKIEW TIDGAIVGQG RTDGRYSAYE PQYLGNDEYN

301  VTLAIAGLTL EDTTKIYNLR ASNELGLTDY QVRISSSSKP PSSSLDVAAI

351  VGIVVAVAVL VLVVLLIVFA RATGRWCFGG KSIKTPTNET SDTESADIKA

401  TSTATATTTM GGVGVSAEEE ETVNEQESPQ EQQQQQQKKA KRLPAFAAAI

451  LRRFNEKDSR KYKDNQESLN IVEGSVQEIP ATNNAIDGND NEPKAIVWQS

501  TSPVWTFK
```

Domain structure
  1–20   Signal sequence
  44–106   Ig-1
  143–215   Ig-2
  256–310   Ig-3
  347–370   Transmembrane domain

Database accession numbers
	PIR	SWISSPROT	EMBL/GENBANK
Drosophila	A33378	P15278	M27813

Alternative forms
A variant with a short cytoplasmic domain is produced by alternative splicing.

References
[1] Grenningloh, G. et al. (1990) Cold Spring Harbor Symp. Quant. Biol. 55, 327–340
[2] Hortsch, M. and Goodman, C.S. (1991) Annu. Rev. Cell Biol. 505–557.
[3] Snow, P.M. et al. (1989) Cell 59, 313–323.

ICAM-1 (intercellular adhesion molecule-1) CD54

Family
Immunoglobulin superfamily, C2 subset.

Cellular distribution
ICAM-1 is basally expressed in significant amounts on a limited number of cell types, including monocytes and endothelial cells. It is widely inducible, or upregulated, on many cells including B and T lymphocytes, thymocytes, dendritic cells, endothelial cells, fibroblasts, keratinocytes, chondrocytes and epithelial cells [1].

Function
ICAM-1 is important in mediating immune and inflammatory responses. ICAM-1 mediates the adhesion of T cells with antigen-presenting cells or target cells and is also involved in T cell/T cell and T cell/B cell interactions. Binding of ICAM-1 has a co-stimulatory effect on T cell activation [2]. ICAM-1 is important in the adhesion of monocytes, lymphocytes and neutrophils to activated endothelium [reviewed in refs 3,4]. It is a receptor for human rhinoviruses [5] and *P.falciparum* infected erythrocytes [6]. Expression of ICAM-1 has been implicated in tumour metastasis [7].

Regulation of expression
Expression of ICAM-1 can be induced or upregulated severalfold by IFNγ, IL1β, TNFα and LPS [1].

Ligands
LFA-1 (CD18/CD11a) [8], MAC-1 (CD18/CD11b) [9], CD43 [10].
Adhesion is calcium-dependent.

Gene structure

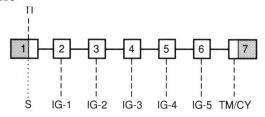

Gene location
19.

Structure [11]

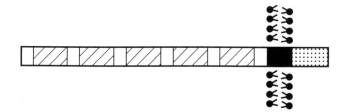

ICAM-1

The LFA-1 binding site is within Ig-domain 1. The Mac-1 binding site is within Ig-domain 3.

Molecular weights
Polypeptide   55 215
SDS PAGE   90 000–115 000

Amino acid sequence (from HL-60) [12]

```
  1  MAPSSPRPAL PALLVLLGAL FPGPGNAQTS VSPSKVILPR GGSVLVTCST

 51  SCDQPKLLGI ETPLPKKELL LPGNNRKVYE LSNVQEDSQP MCYSNCPDGO

101  STAKTFLTVY WTPERVELAP LPSWQPVGKN LTLRCQVEGG APRANLTVVL

151  LRGEKELKRE PAVGEPAEVT TTVLVRRDHH GANFSCRTEL DLRPQGLELF

201  ENTSAPYQLQ TFVLPATPPQ LVSPRVLEVD TQGTVVCSLD GLFPVSEAQV

251  HLALGDQRLN PTVTYGNDSF SAKASVSVTA EDEGTQRLTC AVIIGNQSQE

301  TLQTVTIYSF PAPNVILTKP EVSEGTEVTV KCEAHPRAKV TLNGVPAQPL

351  GPRAQLLLKA TPEDNGRSFS CSATLEVAGQ LIHKNQTREL RVLYGPRLDE

401  RDCPGNWTWP ENSQQTPMCQ AWGNPLPELK CLKDGTFPLP IGESVTVTRD

451  LEGTYLCRAR STQGEVTREV TVNVLSPRYE IVIITVVAAA VIMGTAGLST

501  YLYNRQRKIK KYRLQQAQKG TPMKPNTQAT PP
```

Domain structure
1–25   Signal sequence
41–100   Ig-1
128–190 Ig-2
230–294 Ig-3
325–375 Ig-4
413–461 Ig-5
481–503 Transmembrane domain
152–154 Cell attachment site (potential)

Eight potential N-linked glycosylation sites.

Database accession numbers

	PIR	SWISSPROT	EMBL/GENBANK	REFERENCE
Human	S00573	P05362	X06990	12
Mouse	S06015	P13597	X52264/X16624	13
Rat	S21765	Q00238	X16624 D000913	14

Alternative forms
Soluble forms detected in sera [15].

References
[1] Dustin, M.L. et al. (1986) J. Immunol. 137, 245–254.
[2] Van Seventer, G.A. et al. (1990) J. Immunol. 144, 4579–4586.
[3] **Springer, T.A. et al. (1990) Nature 346, 425–434.**
[4] Wawryk, S.O. et al. (1989) Immunol. Rev. 108, 135–161.
[5] Staunton, D.E. et al. (1990) Cell 61, 243–254.
[6] Berendt, A.R. et al. (1991) Cell 68, 71–87.
[7] Johnson, J.P. et al. (1989) Proc. Natl Acad. Sci.USA 86, 641–644.
[8] Harning, R. et al. (1991) Cancer Res. 51, 5003–5005.
[9] Marlin, S.D. et al. (1987) Cell 51, 813–819.
[10] Diamond, M.S. et al. (1991) Cell 65, 961–971.
[11] Staunton, D.E. et al. (1988) Cell 52, 925–933.
[12] Simmons, D. et al. (1988) Nature 331, 624–627.
[13] Ballantyne, C.M. et al. (1989) Nucleic Acids Res. 17, 5853.
[14] Kita, Y. et al. (1992) Biochim. Biophys. Acta 1131, 108–110.
[15] Rosenstein, Y. et al. (1991) Nature 354, 233–235.

ICAM-2
inter-cellular adhesion molecule-2

Family
Immunoglobulin superfamily, C2 subset.

Cellular distribution
Endothelial cells, subpopulation of lymphocytes, monocytes, splenic sinusoids, dendritic cells [1].

Function
ICAM-2 mediated adhesion can provide a co-stimulatory signal for T cell activation and may therefore be important where antigen-presenting cells express little or no ICAM-1 [2]. The high levels of ICAM-2 present on resting endothelial cells suggest it is the major ligand for LFA-1 and as such may have a role in normal lymphocyte circulation [1].

Regulation of expression
Unlike ICAM-1, ICAM-2 is constitutively expressed on endothelial cells and T cells.

Ligand
LFA-1 (CD11a/CD18, β_2,α_L integrin). Adhesion is calcium-dependent.

Gene structure

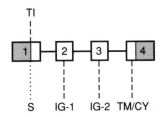

Gene location
17q 23–25

Structure [3]

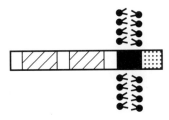

The Ig-like domains are 35% homologous to the two N-terminal domains of ICAM-1.

ICAM-2

Molecular weights
Polypeptide 28 393
SDS PAGE 55 000–65 000

Amino acid sequence [3]

```
                                  ↓                            *
  1   MSSFGYRTLT VALFTLICCP GSDEKVFEVH VRPKKLAVEP KGSLEVNCST
                                       *
 51   TCNQPEVGGL ETSLNKILLD EQAQWKHYLV SNISHDTVLQ CHFTCSGKQE
                 *
101   SMNSNVSVYQ PPRQVILTLQ PTLVAVGKSF TIECRVPTVE PLDSLTLFLF
           *                     *          *
151   RGNETLHYET FGKAAPAPQE ATATFNSTAD REDGHRNFSC LAVLDLMSRG

201   GNIFHKHSAP KMLEIYEPVS DSQMVIIVTV VSVLLSLFVT SVLLCFIFGQ

251   HLRQQRMGTY GVRAAWRRLP QAFRP
```

Domain structure
  1–21   Signal sequence
 42–99   Ig-1
127–194 Ig-2
224–248 Transmembrane domain

Six potential N-linked glycosylation sites.

Database accession numbers
	PIR	SWISSPROT	EMBL/GENBANK	REFERENCE
Human	S03967	P13598	X15606	
Mouse			X65490	4

References
[1] de Fougerolles, A.R. et al. (1991) J. Exp. Med. 174, 253–267.
[2] Damle, N.K. et al. (1992) J. Immunol. 148, 665–671.
[3] Staunton, D.E. et al. (1989) Nature 339, 61–64.
[4] Xu, H. et al. (1992) J. Immunol. 149, 2650–2655.

Integrin α6β4

α_E β4, TSP 180 antigens (mouse)

Family
Integrin (dimer of α6 and β4 subunits).

Cellular distribution
Epithelia, Schwann cells, some tumour cells [1,2], some endothelia and neuronal cells [3].

Function
Laminin receptor [4]. The localization of α6β4 on the basal surface of polarized epithelial cells [5] and in human hemidesmosomes [6] suggests that it may be involved in the adhesion of cells to the extracellular matrix. It may, therefore, play a role in the establishment of adhesion complexes during wound healing and in the maintenance of adhesion of epithelial sheets to underlying connective tissue in unwounded areas [7]. α6β4 has also been implicated in NK and T cell mediated cytotoxicity [8].

Regulation of expression
Thought to be constitutively expressed; however, the laminin binding function requires activation.

Ligand
Laminin (E8 fragment) [9,4]. Possibly binds epiligrin.

Gene structure
Unknown.

Gene location
α6 chromosome 2.
β4 chromosome 17q11-qter [10].

Structure

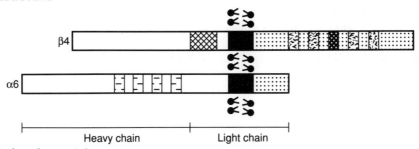

Molecular weights
Polypeptide   117 263 (α6); 198 011 (β4)
SDS PAGE   150 000 (α6); 205 000 (β4)
               α6 subunit cleaved into 125 000/25 000 heavy and light chains.
               β4 migrates as polypeptides of 205 000, 180 000, 150 000 [1]

Amino acid sequence (α6)
See entry for VLA-6.

Integrin α6β4

Amino acid sequence (β4) (from retinal pigment epithelial cells) [11]

```
   1  MAGPRPSPWA RLLLAALISV SLSGTLANRC KKAPVKSCTE CVRVDKDCAY
  51  CTDEMFRDRR CNTQAELLAA GCQRESIVVM ESSFQITEET QIDTTLRRSQ
 101  MSPQGLRVRL RPGEERHFEL EVFEPLESPV DLYILMDFSN SMSDDLDNLK
 151  KMGQNLARVL SQLTSDYTIG FGKFVDKVSV PQTDMRPEKL KEPWPNSDPP
 201  FSFKNVISLT EDVDEFRNKL QGERISGNLD APEGGFDAIL QTAVCTRDIG
 251  WRPDSTHLLV FSTESAFHYE ADGANVLAGI MSRNDERCHL DTTGTYTQYR
 301  TQDYPSVPTL VRLLAKHNII PIFAVTNYSY SYYEKLHTYF PVSSLGVLQE
 351  DSSNIVELLE EAFNRIRSNL DIRALDSPRG LRTEVTSKMF QKTRTGSFHI
 401  RRGEVGIYQV QLRALEHVDG THVCQLPEDQ KGNIHLKPSF SDGLKMDAGI
 451  ICDVCTCELQ KEVRSARCSF NGDFVCGQCV CSEGWSGQTC NCSTGSLSDI
 501  QPCLREGEDK PCSGRGECQC GHCVCYGEGR YEGQFCEYDN FQCPRTSGFL
 551  CNDRGRCSMG QCVCEPGWTG PSCDCPLSNA TCIDSNGGIC NGRGHCECGR
 601  CHCHQQSLYT DTICEINYSA IHPGLCEDLR SCVQCQAWGT GEKKGRTCEE
 651  CNFKVKMVDE LKRAEEVVVR CSFRDEDDDC TYSYTMEGDG APGPNSTVLV
 701  HKKKDCPPGS FWWLIPLLLL LLPLLALLLL LCWKYCACCK ACLALLPCCN
 751  RGHMVGFKED HYMLRENLMA SDHLDTPMLR SGNLKGRDVV RWKVTNNMQR
 801  PGFATHAASI NPTELVPYGL SLRLARLCTE NLLKPDTREC AQLRQEVEEN
 851  LNEVYRQISG VHKLQQTKFR QQPNAGKKQD HTIVDTVLMA PRSAKPALLK
 901  LTEKQVEQRA FHDLKVAPGY YTLTADQDAR GMVEFQEGVE LVDVRVPLFI
 951  RPEDDDEKQL LVEAIDVPAG TATLGRRLVN ITIIKEQARD VVSFEQPEFS
1001  VSRGDQVARI PVIRRVLDGG KSQVSYRTQD GTAQGNRDYI PVEGELLFQP
1051  GEAWKELQVK LLELQEVDSL LRGRQVRRFH VQLSNPKFGA HLGQPHSTTI
1101  IIRDPDELDR SFTSQMLSSQ PPPHGDLGAP QNPNAKAAGS RKIHFNWLPP
```

```
1151  SGKPMGYRVK  YWIQGDSESE  AHLLDSKVPS  VELTNLYPYC  DYEMKVCAYG
1201  AQGEGPYSSL  VSCRTHQEVP  SEPGRLAFNV  VSSTVTQLSW  AEPAETNGEI
1251  TAYEVCYGLV  NDDNRPIGPM  KKVLVDNPKN  RMLLIENLRE  SQPYRYTVKA
1301  RNGAGWGPER  EAIINLATQP  KRPMSIPIIP  DIPIVDAQSG  EDYDSFLMYS
1351  DDVLRSPSGS  QRPSVSDDTE  HLVNGRMDFA  FPGSTNSLHR  MTTTSAAAYG
1401  THLSPHVPHR  VLSTSSTLTR  DYNSLTRSEH  SHSTTLPRDY  STLTSVSSHD
1451  SRLTAGVPDT  PTRLVFSALG  PTSLRVSWQE  PRCERPLQGY  SVEYQLLNGG
                                     *
1501  ELHRLNIPNP  AQTSVVVEDL  LPNHSYVFRV  RAQSQEGWGR  EREGVITIES
1551  QVHPQSPLCP  LPGSAFTLST  PSAPGPLVFT  ALSPDSLQLS  WERPRRPNGD
1601  IVGYLVTCEM  AQGGGPATAF  RVDGDSPESR  LTVPGLSENV  PYKFKVQART
1651  TEGFGPEREG  IITIESQDGG  PFPQLGSRAG  LFQHPLQSEY  SSITTTHTSA
1701  TEPFLVDGPT  LGAQHLEAGG  SLTRHVTQEF  VSRTLTTSGT  LSTHMDQQFF
1751  QT
```

Domain structure (β4)

1–27	Signal sequence
28–710	Extracellular domain
711–733	Transmembrane domain
456–502	Cysteine-rich repeat
503–542	Cysteine-rich repeat
543–581	Cysteine-rich repeat
582–619	Cysteine-rich repeat
734–1752	Cytoplasmic domain
1125–1214	FN-III
1219–1314	FN-III
1456–1546	FN-III
1570–1662	FN-III

Eight potential N-linked glycosylation sites (five in extracellular domain).

Tyr_{1690} in the cytoplasmic domain may be a candidate site for phosphorylation since the surrounding sequence shows similarity to the EGF and insulin receptors.

Database accession numbers (β4)

	PIR	SWISSPROT	EMBL/GENBANK	REFERENCE
Human		P16144	X51841	*11*
Mouse			X58254	

Alternative forms

Multiple mRNA forms (β4) detected by reverse transcriptase-PCR have been identified in both normal and transformed epithelial cell types. In the cytoplasmic tail, an in-frame insertion of 70 amino acids has been identified between Thr_{1369} and Glu_{1440} after nucleotide 4292 [12]. Similarly, a 53 amino acid insertion in the cytoplasmic tail has been detected at nucleotide 4744 [13]. Tamura et al. (1990) have also identified a cDNA containing an additional 49 bp in the 5' untranslated region, 9 bp upstream from the ATG initiation codon. This insertion may have some regulatory significance.

At the protein level, three forms of the β4 chain have been identified (see above). It is assumed that the three forms arise from proteolytic cleavage of the 205 kD band and that each form can associate independently with the α6 chain [2]. However, the functional significance is unknown at present.

References
[1] **Kajiji, S. et al. (1989) EMBO J. 8, 673–680.**
[2] Hemler, M.E. et al. (1989) J. Biol. Chem. 264, 6529–6535.
[3] Quaranta and Jones (1991) Trends Cell Biol. 1, 2–4.
[4] Lee, E.C. et al. (1992) J. Cell Biol. 117, 671–678.
[5] Sonnenberg, A. et al. (1990) J. Cell Sci. 96, 207–217.
[6] Sonnenberg, A. et al. (1991) J. Cell Biol. 113, 907–917.
[7] Kurpakus, M.A. et al. (1991) J. Cell Biol. 115, 1737–1750.
[8] Phillips, J.H. et al. (1991) J. Exp. Med. 174, 1571–1581.
[9] Lotz, M.M. et al. (1990) Cell Regulation 1, 249–257.
[10] Hogervorst, F. et al. (1991) Eur. J. Biochem. 199, 425–433.
[11] **Suzuki, S. and Naitoh, Y. (1990) EMBO J. 9, 757–763.**
[12] Tamura, R.N. et al. (1990) J. Cell Biol. 111, 1593–1604.
[13] Hogervorst, F. et al. (1990) EMBO J. 9, 765–770.

Integrin α7β1

α7 subunit also known as H36

Family
β1 integrin (dimer of α7 and β1 subunit).

Cellular distribution
Melanoma cells [1,2] and rat and mouse myoblasts [3], skeletal and cardiac muscle [4].

Function
Cell adhesion to laminin [1].

Regulation of expression
Expression is developmentally regulated during skeletal myogenesis [4]. Expression of the α7 subunit is regulated both early in the development of the myogenic lineage and later during terminal differentiation. Expression of the β1 subunit does not coincide fully with that of α7 during myogenesis suggesting that the α7 chain may interact with an as yet unidentified β subunit. Expression may also be associated with malignant transformation and therefore α7β1 may play a role in influencing metastatic potential.

Ligand
Laminin (binds to E8 region) [1].

Gene structure
Unknown.

Gene location
Unknown.

Structure

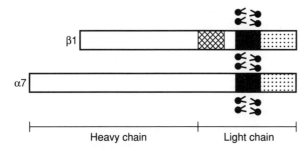

Heavy chain | Light chain

Molecular weights
Polypeptide   120 957 (α7)
SDS-PAGE   120 000–125 000 (α7) cleaved into heavy and light chain of approximately 100 000 and 30 000.

Amino acid sequence (α7) (incomplete sequence from rat myotubes) [4]

```
   1  FNLDVMGAIR KEGEPGSLFG FSVALHRQLQ PRPQSWLLVG APQALDSYPD
  51  SRQIAHGRPL CLSLSEETD  CYRVDIDRGA NVQKESKENQ WLGVSVRPRE
 101  SGGKVVTCAH RYESRQEVDQ VLETRDVIGR CFVLSQDLAI RDELDGGEWK
 151  FCEGRPQGHE QFGFCQQGTA ATFSPDSHYL IFGAPGTYNW KGTARVELCA
 201  QGSSDLAQVD DGPYEAGGEK DQDPRPSPVP ANSYLGFSID SGKGLMRSEE
 251  LSFVAGAPRA NHKGAVVILR KDSASRLIPE VVLSGERLTS GFGYSLAVTD
 301  LNSDGWADLI VGAPYFFERQ EELGGAVYVY MNQGGHWADI SPLRLCGSPH
 351  SMFGISLAVL GDLNQDGFPD IAVGAPFDGD GKVFIYHGSS LGVVTKPSQV
 401  LEGEAVGIKS FGYSLSGGLD VDGNHYPDLL VGSLADTAAL FRARPVLHVS
 451  QEIFIDPRAI DLEQPNCADG RLVCVHVKVC FSYVAVPSSY SPIVVLDYVL
 501  DGDTDRRLRG QAPRVTFPGR GPDDLKHQSS GTVSLKHQHD RVCGDTCVPA
 551  AGKRKDKLRA IVVTLSYGLQ TPRLRRQAPD QGLPLVAGIL NAHQPSTQRT
 601  EIHFLKQGCG DDKICQSNLQ LVQAQFCSRI SDTEFQALPM DLDGTALFAH
 651  GGQPFIGLEL TVTNLPSDPA RPQADGDDAH EAQLLATLPA SLRYSGVRTL
 701  DSVEKPLCLS NENASHVECE LGNPMKRGTQ VTFYLILSTS GITIETTELK
 751  VELLLATISE QDLHPVSVRA HVFIELPLSI SGVATPQQLF FSGKVKGESA
 801  MRSEWDEGSK VKYEVTVSNQ GQSLNTLGSA FLNIMWPHEI ANGKWLLYPM
 851  RVELEGGQGP EKKGICSPRP NILHLDVDSR DRRRRELGQP EPQEPPEKVE
 901  PSTSWWPVSS AEKRNVTLDC PGTAKCVVFS CPLYSFDRAA VLHVWGRLWN
 951  STFLEEYMSV KSLEVIVRAN ITVKSSIKNL LLRDASTVIP VMVYLDPVAV
1001  VAEGVPWWVI LLAVLAGLLV LALLVLLLWK LGFFKRAKHP EATVPQYHAV
1051  KILREDRQQF KEEKTGTIQR SNWGNSQWEG SDAHPILAAD WHPELGPDGH
1101  PVSVTA
```

Integrin α7β1

The N-terminal sequence shown here is almost identical to amino acid sequence data obtained from two human melanoma cell lines [1].

Domain structure

1–1006	Extracellular domain
1007–1029	Transmembrane domain
1030–1106	Cytoplasmic domain
300–308	Cation binding site
361–370	Cation binding site
420–428	Cation binding site

The extracellular domain contains two potential protease cleavage sites at position 575 (RRQ) and position 882 (RRRRE).

There are four potential N-linked glycosylation sites.

Amino acid sequence (β1)

See entry for VLA-1.

Database accession numbers (α7)

	PIR	SWISSPROT	EMBL/GENBANK
Rat			X65036

References

[1] **Kramer R.H. et al. (1991) Cell Regulation 2, 805–817.**
[2] Kramer R.H. et al. (1989) J. Biol. Chem. 264, 15642–15649.
[3] von der Mark H. et al. (1991) J. Biol. Chem. 266, 23593–23601.
[4] Song, W.K. et al. (1992) J. Cell Biol. 643–657.

L1

NILE (rat homologue)

Family
Immunoglobulin superfamily, C2 subset. L1 is structurally related to Ng-CAM and shares 40% amino acid identity.

Cellular distribution
L1 is expressed on post-mitotic, premigratory neuronal cell bodies. On post-migratory neurons, L1 is localized predominantly to axons. L1 is also present on Schwann cells and on a subpopulation of epithelial cells [reviewed in ref.1].

Function
L1 mediates neuron/neuron and neuron/glial cell interactions. It is implicated in neuronal and epithelial cell migration, neurite fasciculation and outgrowth [reviewed in ref.1]. The binding of antibodies to L1 can lead to changes in levels of intracellular IP2, IP3 and calcium. There is evidence that the rise in calcium involves signalling involving a G-protein [2]. A protein kinase is associated with L1 and phosphorylates the cytoplasmic domain [3].

Regulation of expression
L1 is expressed on post-mitotic neurons during the period of neurite extension [4]. Expression of L1, as determined by immunocytochemistry, is downregulated on mature neurons, but this may reflect obscuring of the antigen [5].

Ligands
L1 binds to axonin-1 on neurons [6] and to an unknown ligand on glia. L1 interacts functionally with NCAM in the plane of the membrane, inducing a synergistic increase in adhesion [7]. Adhesion is calcium-independent.

Gene location
X-chromosome.

Structure

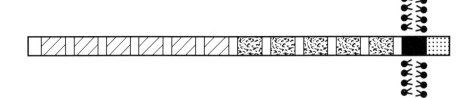

Molecular weights
Polypeptide   138 735
SDS PAGE   200 000 (predominant), 180 000, 140 000, 80 000 and 30 000

L1

Amino acid sequence (from 8 day postnatal mouse brain) [9]

```
   1  MVVMLRYVWP LLLCSPCLLI QIPDEYKGHH VLEPPVITEQ SPRRLVVFPT
                          ↓                                *
  51  DDISLKCEAR GRPQVEFRWT KDGIHFKPKE ELGVVVHEAP YSGSFTIEGN
 101  NSFAQRFQGI YRCYASNKLG TAMSHEIQLV AEGAPKWPKE TVKPVEVEEG
 151  ESVVLPCNPP PSAAPPRIYW MNSKIFDIKQ DERVSMGQNG DLYFANVLTS
        *                                              *
 201  DNHSDYICNA HFPGTRTIIQ KEPIDLRVKP TNSMIDRKPR LLFPTNSSSR
                                                        *
 251  LVALQGQSLI LECIAEGFPT PTIKWLHPSD PMPTDRVIYQ NHNKTLQLLN
 301  VGEEDDGEYT CLAENSLGSA RHAYYVTVEA APYWLQKPQS HLYGPGETAR
 351  LDCQVQGRPQ PEITWRINGM SMETVNKDQK YRIEQGSLIL SNVQPTDTMV
                                         *
 401  TQCEARNQHG LLLANAYIYV VQLPARILTK DNQTYMAVEG STAYLLCKAF
                                *          *
 451  GAPVPSVQWL DEEGTTVLQD ERFFPYANGT LSIRDLQAND TGRYFCQAAN
        *
 501  DQNNVTILAN LQVKEATQIT QGPRSAIEKK GARVTFTCQA SFDPSLQASI
                                                  *
 551  TWRGDGRDLQ ERGDSDKYFI EDGKLVIQSL DYSDQGNYSC VASTELDEVE
 601  SRAQLLVVGS PGPVPHLELS DRHLLKQSQV HLSWSPAEDH NSPIEKYDIE
                                *
 651  FEDKEMAPEK WFSLGKVPGN QTSTTLKLSP YVHYTFRVTA INKYGPGEPS
                                *
 701  PVSESVVTPE AAPEKNPVDV RGEGNETNNM VITWKPLRWM DWNAPQIQYR
                                *
 751  VQWRPQGKQE TWRKQTVSDP FLVVSNTSTF VPYEIKVQAV NNQGKGPEPQ
                                *                        *
 801  VTIGYSGEDY PQVSPELEDI TIFNSSTVLV RWRPVDLAQV KGHLKGYNVT
                                          *
 851  YWWKGSQRKH SKRHIHKSHI VVPANTTSAI LSGLRPYSSY HVEVQAFNGR
 901  GLGPASEWTF STPEGVPGHP EALHLECQSD TSLLLHWQPP LSHNGVLTGY
                                *          *
 951  LLSYHPVEGE SKEQLFFNLS DPELRTHNLT NLNPDLQYRF QLQATTQQGG
                                *          *
1001  PGEAIVREGG TMALFGKPDF GNISATAGEN YSVVSWVPRK GQCNFRFHIL
                                          *
1051  FKALPEGKVS PDHQPQPQYV SYNQSSYTQW NLQPDTKYEI HLIKEKVLLH
           *
1101  HLDVKTNGTG PVRVSTTGSF ASEGWFIAFV SAIILLLLIL LILCFIKRSK
1151  GGKYSVKDKE DTQVDSEARP MKDETFGEYR SLESDNEEKA FGSSQPSLNG
1201  DIKPLGSDDS LADYGGSVDV QFNEDGSFIG QYSGKKEKEA AGGNDSSGAT
1251  SPINPAVALE
```

L1

Domain structure

1–19	Signal sequence
50–117	Ig-1
150–212	Ig-2
256–315	Ig-3
346–407	Ig-4
440–503	Ig-5
531–594	Ig-6
624–709	FN-III
734–815	FN-III
831–917	FN-III
936–1018	FN-III
1036–1111	FN-III
1124–1146	Transmembrane domain

Potential cell attachment sites 553–555, 562–564. Six potential N-linked glycosylation sites.

Twenty-one potential N-linked glycosylation sites.

Database accession numbers

	PIR	SWISSPROT	EMBL/GENBANK
Mouse	S05479	P11627	X12875

Rat The rat homologue of L1 has been isolated [5].

Human The human homologue of L1 has been isolated [10].

Alternative forms

L1 is synthesized as transmembrane protein of M_r 200 000. Fragments of M_r 180 000/140 000 and M_r 30 000/80 000 are generated by proteolytic cleavage. The M_r 180 000/140 000 fragment consists of the six amino-terminal Ig-like domains, the first two fibronectin repeats and a portion of the third. The M_r 30 000/80 000 fragment consists of the remaining fibronectin repeats, together with the transmembrane and cytoplasmic domains. The M_r 180 000/140 000 fragment is non-covalently associated with the membrane bound fragment [8]. The cleavage site is between Arg 863 and His 864. Two variants differing in the cytoplasmic domain have been detected in the rat [5] and human [10] produced by alternative splicing.

References

[1] **Schachner, M. (1990) Ciba Foundation Symposia 145, 156–172.**
[2] Schuch, U. et al. (1989) Neuron 3, 13–20.
[3] Sadoul, R. et al. (1989) J. Neurochem. 53, 1471–1478.
[4] Martini, R. and Schachner, M. (1988) J. Cell Biol. 100, 1753–1746.
[5] Prince, J.T. et al. (1991). J. Neurosci. Res. 30, 567–581.
[6] Kuhn, T.B. et al. (1991) J. Cell Biol. 115, 1113–1126.
[7] Kadmon, G. et al. (1990) J. Cell Biol. 110, 209–218.
[8] Sadoul, K. et al. (1988) J. Neurochem. 50, 510–521.
[9] Moos, M. et al. (1988) Nature 334, 701–703.
[10] Reid, R.A. and Hemperley, J.J. (1992) J. Mol. Neurosci 3, 127–135.

Leukocyte adhesion receptor p150,95

CD11c/CD18, αXβ2, Leu M5

Other names
Complement receptor type 4 (CR4), LeuCAMc.

Family
β2 integrin (dimer of αX and β2 subunits).

Cellular distribution

Macrophages, monocytes, granulocytes, activated T and B lymphocytes, NK cells, dendritic cells, hairy leukaemia cells [1], B cell chronic lymphocytic leukaemias [2] and microglia [3].

Function

Monocyte/granulocyte adhesion to endothelium during inflammatory responses [4]. Fibrinogen binding [5]. Neutrophil adhesion to serum-coated surfaces. Chemotaxis and adhesion of peripheral blood monocytes [6]. Involved in B cell activation [7]. May play a role in CTL mediated killing [8]. p150,95 is also involved in the respiratory burst in granulocytes [9].

Regulation of expression

Inflammatory mediators induce rapid mobilization of a large intracellular pool of p150,95 to the cell surface [10]. Can be induced in purified human B cells by phorbol ester treatment. *In vivo*, p150,95 is upregulated on differentiation of peripheral blood monocytes into tissue macrophages following extravasation which is often associated with a decrease in MAC-1 expression [1]. p150,95 is also upregulated on activation of B lymphocytes [11].

Ligand
Fibrinogen via a GPRP sequence [7,12].
p150,95 has also been reported to bind iC3b [13] and denatured albumin [14].
The ligand on endothelial cells is not known.

Gene structure

The αX gene spans 30 kb, and consists of 31 exons grouped in five clusters [15,16]. Each of the three divalent cation binding sites is encoded by a separate exon. Both the signal peptide and transmembrane domain are split into two exons. The I-domain is distributed over four exons.

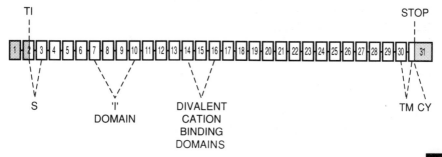

Leukocyte adhesion receptor p150,95

Gene location and size
16p11–13.1; 30 kb.[17]

Structure

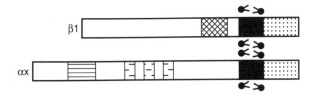

Molecular weights
Polypeptide   125 908 (αX); 82 572 (β2)
SDS PAGE   150 000 (αX); 95 000 (β2)

Amino acid sequence (αX) (from PMA-stimulated HL-60 cells) [18].

```
  1  MTRTRAALLL  FTALATSLGF  NLDTEELTAF  RVDSAGFGDS  VVQYANSWVV
                     *                       *
 51  VGAPQKITAA  NQTGGLYQCG  YSTGACEPIG  LQVPPEAVNM  SLGLSLASTT
101  SPSQLLACGP  TVHHECGRNM  YLTGLCFLLG  PTQLTQRLPV  SRQECPRQEQ
151  DIVFLIDGSG  SISSRNFATM  MNFVRAVISQ  FQRPSTQFSL  MQFSNKFQTH
201  FTFEEFRRTS  NPLSLLASVH  QLQGFTYTAT  AIQNVVHRLF  HASYGARRDA
251  TKILIVITDG  KKEGDSLDYK  DVIPMADAAG  IIRYAIGVGL  AFQNRNSWKE
301  LNDIASKPSQ  EHIFKVEDFD  ALKDIQNQLK  EKIFAIEGTE  TTSSSSFELE
                                             *           *
351  MAQEGFSAVF  TPDGPVLGAV  GSFTWSGGAF  LYPPNMSPTF  INMSQENVDM
401  RDSYLGYSTE  LALWKGVQSL  VLGAPRYQHT  GKAVIFTQVS  RQWRMKAEVT
451  GTQIGSYFGA  SLCSVDVDTD  GSTDLVLIGA  PHYYEQTRGG  QVSVCPLPRG
501  WRRWWCDAVL  YGEQGHPWGR  FGAALTVLGD  VNGDKLTDVV  IGAPGEEENR
551  GAVYLFHGVL  GPSISPSHSQ  RIAGSQLSSR  LQYFGQALSG  GQDLTQDGLV
601  DLAVGARGQV  LLLRTRPVLW  VGVSMQFIPA  EIPRSAFECR  EQVVSEQTLV
                                                             *
651  QSNICLYIDK  RSKNLLGSRD  LQSSVTLDLA  LDPGRLSPRA  TFQETKNRSL
                                        *
701  SRVRVLGLKA  HCENFNLLLP  SCVEDSVTPI  TLRLNFTLVG  KPLLAFRNLR
```

Leukocyte adhesion receptor p150,95

```
 751   PMLAADAQRY FTASLPFEKN CGADHICQDN LGISFSFPGL KSLLVGSNLE

 801   LNAEVMVWND GEDSYGTTIT FSHPAGLSYR YVAEGQKQGQ LRSLHLTCDS
                                                           *
 851   APVGSQGTWS TSCRINHLIF RGGAQITFLA TFDVSPKAVL GDRLLLTANV
                *                                  *
 901   SSENNTPRTS KTTFQLELPV KYAVYTVVSS HEQFTKYLNF SESEEKESHV

 951   AMHRYQVNNL GQRDLPVSIN FWVPVELNQE AVWMDVEVSH PQNPSLRCSS
                                                           *
1001   EKIAPPASDF LAHIQKNPVL DCSIAGCLRF RCDVPSFSVQ EELDFTLKGN

1051   LSFGWVRQIL QKKVSVVSVA EITFDTSVYS QLPGQEAFMR AQTTTVLEKY

1101   KVHNPTPLIV GSSIGGLLLL ALITAVLYKV GFFKRQYKEM MEEANGQIAP

1151   ENGTQTPSPP SEK
```

Domain structure
Seven homologous internal repeats in extracellular domain and an additional I-domain (187 amino acids) between repeats II and III. There are also three divalent cation binding sites in the αX subunit.

- 1–19 Signal sequence
- 20–1107 Extracellular domain
- 443–638 3 tandem repeats of approximately 60 amino acids containing divalent cation binding sites.
- 1108–1128 Transmembrane domain
- 1129–1163 Cytoplasmic domain
- 146–338 I-domain

The αX subunit is constitutively phosphorylated.

Eight potential N-linked glycosylation sites.

Amino acid sequence (β2)
See entry for LFA-1.

Database accession numbers (αX)

	PIR	SWISSPROT	EMBL/GENBANK	REFERENCE
Human	S00864		Y00093	
		P20702		18

References
1. **Larson, R.S. and Springer, T.A. (1990) Immunol. Rev. 114, 181–217.**
2. De la Hera, A. et al. (1988) Eur. J. Immunol. 18, 1131–1134.
3. Aklyama, H. and McGear P.L. (1990) J. Neuroimmunol. 30, 81–93.
4. Stacker, S.A. and Springer, T.A. (1991) J. Immunol. 146, 648–655.

5 Myones, B.L. et al. (1988) J. Clin. Invest. 82, 640–651.
6 Te Velde, A.A. et al. (1987) Immunology 61, 261–267.
7 Postigo, A. et al. (1991) J. Exp. Med. 174, 1313–1322.
8 Keizer, G.D. et al. (1987) J. Immunol. 138, 3130–3136.
9 Berton, G. et al. (1992) J. Cell Biol. 118, 1007–1117.
10 Anderson, D.C. and Springer, T.A. (1987) Annu. Rev. Med. 38, 175–194.
11 **Sanchez-Madrid, F. and Corbi, A.L. (1993) Seminars Cell Biol. 3, 199–210.**
12 Loike, J.D. et al. (1991) Proc. Natl Acad. Sci. USA 88, 1044–1048.
13 Micklem, K.J. and Sim, R.B. (1985) Biochem. J. 231, 233–236.
14 Davis, G. (1992) Exp. Cell Res. 200, 242–252.
15 Corbi, A.L. et al. (1990) J. Biol. Chem. 265, 2782–2788.
16 Noti, J.D. et al. (1992) DNA Cell Biol. 11, 123–138.
17 Corbi, A.L. et al. (1988) J. Exp. Med. 167, 1597–1607.
18 Corbi, A.L. et al. (1987) EMBO J. 6, 4023–4028.

LFA-1

(leukocyte function associated molecule-1) CD11a/CD18, integrin αLβ2

Family
β2 integrin (dimer of αL and β2 subunits).

Cellular distribution
Lymphocytes, neutrophils, monocytes and macrophages.

Function
The proposed functions of LFA-1 have recently been reviewed [1].

LFA-1 has a key role in mediating leukocyte adhesion to endothelium during inflammatory responses through binding to ICAM-1. It is also involved in most immune phenomena involving T lymphocytes such as adhesion of cytotoxic T cells to their target cells, mixed lymphocyte reactions, antigen-specific and ConA-induced T cell proliferation, and T cell-dependent antibody response. LFA-1 is also involved in homotypic and heterotypic leukocyte interactions between B and T cell lines [reviewed in ref 1].

Signal transduction is thought to occur through the β2 subunit. Ser_{756} is phosphorylated in response to phorbol ester [2]. In vitro, purified protein kinase C is able to phosphorylate the β2 subunit [3]. Deletion of the β2 cytoplasmic domain leads to expression of an inactive receptor whereas deletion of the α subunit cytoplasmic domain has no effect [2]. Cross-linking of LFA-1 on peripheral blood T cells by anti-LFA-1 antibodies or by immobilized ICAM-1 in combination with anti-CD3 antibodies results in an alteration of intracellular Ca^{2+}, increased DNA synthesis and IL2 production [4,20].

Regulation of expression
Mainly constitutively expressed. Functional activity of LFA-1 depends on Mg^{2+} dependent activation rather than on significant quantitative changes on the cell surface and requires conserved sequences in the cytoplasmic domain of the β2 subunit [5]. Activation can be triggered through the TCR/CD3 complex or by treatment with phorbol ester [6,7]. αL chain expression is reduced in lymphoblastoid cell lines transformed by c-myc whereas β2 expression remains unchanged [8].

A defect in β2 expression results in leukocyte adhesion deficiency (LAD) in which patients suffer from recurrent bacterial infections [9].

Ligands
ICAM-1 (via its N-terminal Ig-domain) [10], ICAM-2 [11], ICAM-3 [12].

Gene structure
For the αL gene a genomic clone extending over 32 kb has been identified which encodes over 70% of the residues in the extracellular domain. The nucleotide sequence of six exons shows splice sites at identical positions to those in p150,95, suggesting that the intron/exon organization of the αL subunit may be identical to that of αX [Ref 1 and Larson and Springer, unpublished results].

The β2 gene contains 16 exons spanning 40 kb [13]. Transcription initiates from five sites which may be due to the absence of an upstream TATA box. Five different polyadenylation sites have also been identified.

LFA-1

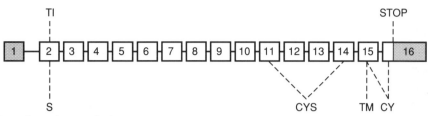

Gene location and size
16p11–13.1; 32 kb (αL) [14]
21q22.3; 40 kb (β2) [15].

Structure

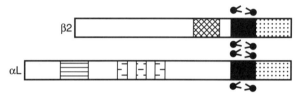

Molecular weights
Polypeptide   126 193 (αL); 82 572 (β2)
SDS PAGE   180 000 (αL); 95 000 (β2)

Amino acid sequence (αL) (from PMA-stimulated HL-60 cells) [16].

```
  1  MKDSCITVMA MALLSGFFFF APASSYNLDV RGARSFSPPR AGRHFGYRVL
 51  QVGNGVIVGA PGEGNSTGSL YQCQSGTGHC LPVTLRGSNY TSKYLGMTLA
101  TDPTDGSILA CDPGLSRTCD QNTYLSGLCY LFRQNLQGPM LQGRPGFQEC
151  IKGNVDLVFL FDGSMSLQPD EFQKILDFMK DVMKKLSNTS YQFAAVQFST
201  SYKTEFDFSD YVKWKDPDAL LKHVKHMLLL TNTFGAINYV ATEVFREELG
251  ARPDATKVLI IITDGEATDS GNIDAAKDII RYIIGIGKHF QTKESQETLH
301  KFASKPASEF VKILDTFEKL KDLFTELQKK IYVIEGTSKQ DLTSFNMELS
351  SSGISADLSR GHAVVGAVGA KDWAGGFLDL KADLQDDTFI GNEPLTPEVR
401  AGYLGYTVTW LPSRQKTSLL ASGAPRYQHM GRVLLFQEPQ GGGHWSQVQT
451  IHGTQIGSYF GGELCGVDVD QDGETELLLI GAPLFYGEQR GGRVFIYQRR
501  QLGFEEVSEL QGDPGYPLGR FGEAITALTD INGDGLVDVA VGAPLEEQGA
551  VYIFNGRHGG LSPQPSQRIE GTQVLSGIQW FGRSIHGVKD LEGDGLADVA
```

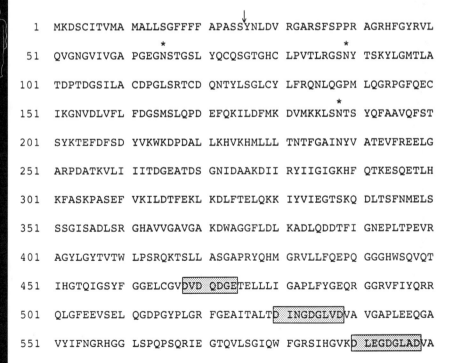

94

LFA-1

```
                                                                   *
 601   VGAESQMIVL SSRPVVDMVT LMSFSPAEIP VHEVECSYST SNKMKEGVNI
                             *
 651   TICFQIKSLY PQFQGRLVAN LTYTLQLDGH RTRRRGLFPG GRHELRRNIA
                                   *    *
 701   VTTSMSCTDF SFHFPVCVQD LISPINVSLN FSLWEEEGTP RDQRAQGKDI

 751   PPILRPSLHS ETWEIPFEKN CGEDKKCEAN LRVSFSPARS RALRLTAFAS

 801   LSVELSLSNL EEDAYWVQLD LHFPPGLSFR KVEMLKPHSQ IPVSCEELPE
                        *                    *              *
 851   ESRLLSRALS CNVSSPIFKA GHSVALQMMF NTLVNSSWGD SVELHANVTC

 901   NNEDSDLLED NSATTIIPIL YPINILIQDQ EDSTLYVSFT PKGPKIHQVK

 951   HMYQVRIQPS IHDHNIPTLE AVVGVPQPPS EGPITHQWSV QMEPPVPCHY

1001   EDLERLPDAA EPCLPGALFR CPVVFRQEIL VQVIGTLELV GEIEASSMFS
                 *          *
1051   LCSSLSISFN SSKHFHLYGS NASLAQVVMK VDVVYEKQML YLYVLSGIGG

1101   LLLLLLIFIV LYKVGFFKRN LKEKMEAGRG VPNGIPAEDS EQLASGQEAG

1151   DPGCLKPLHE KDSESGGGKD
```

Domain structure

The αL subunit contains seven homologous repeats with an additional I-domain inserted between repeats II and III. αL contains three divalent metal cation binding sites of calmodulin EF-hand type. Mg^{2+} is required for cell–cell interactions.

1–25	Signal sequence	
32–79	Repeat	I
82–132	Repeat	II
339–391	Repeat	III
392–446	Repeat	IV
447–508	Repeat	V
509–567	Repeat	VI
568–629	Repeat	VII
170–349	I-domain	
1089–1112	Transmembrane domain	
1113–1170	Cytoplasmic domain	

αL is constitutively phosphorylated.

Twelve potential N-linked glycosylation sites.

LFA-1

Amino acid sequence (β2) (from human tonsil) [18]

```
  1   MLGLRPPLLA LVGLLSLGCV LSQECTKFKV SSCRECIESG PGCTWCQKLN
 51   FTGPGDPDSI RCDTRPQLLM RGCAADDIMD PTSLAETQED HNGGQKQLSP
101   QKVTLYLRPG QAAAFNVTFR RAKGYPIDLY YLMDLSYSML DDLRNVKKLG
151   GDLLRALNEI TESGRIGFGS FVDKTVLPFV NTHPDKLRNP CPNKEKECQP
201   PFAFRHVLKL TNNSNQFQTE VGKQLISGNL DAPEGGLDAM MQVAACPEEI
251   GWRNVTRLLV FATDDGFHFA GDGKLGAILT PNDGRCHLED NLYKRSNEFD
301   YPSVGQLAHK LAENNIQPIF AVTSRMVKTY EKLTEIIPKS AVGELSEDSS
351   NVVHLIKNAY NKLSSRVFLD HNALPDTLKV TYDSFCSNGV THRNQPRGDC
401   DGVQINVPIT FQVKVTATEC IQEQSFVIRA LGFTDIVTVQ VLPQCECRCR
451   DQSRDRSLCH GKGFLECGIC RCDTGYIGKN CECQTQGRSS QELEGSCRKD
501   NNSIICSGLG DCVCGQCLCH TSDVPGKLIY GQYCECDTIN CERYNGQVCG
551   GPGRGLCFCG KCRCHPGFEG SACQCERTTE GCLNPRRVEC SGRGRCRCNV
601   CECHSGYQLP LCQECPGCPS PCGKYISCAE CLKFEKGPFG KNCSAACPGL
651   QLSNNPVKGR TCKERDSEGC WVAYTLEQQD GMDRYLIYVD ESRECVAGPN
701   IAAIVGGTVA GIVLIGILLL VIWKALIHLS DLREYRRFEK EKLKSQWNND
751   NPLFKSATTT VMNPKFAES
```

Domain structure

The β2 subunit contains four cysteine-rich repeats in the extracellular domain.

1–22	Signal sequence
449–496	Cysteine-rich repeat
497–540	Cysteine-rich repeat
541–581	Cysteine-rich repeat
582–617	Cysteine-rich repeat
701–723	Transmembrane domain
724–769	Cytoplasmic domain.

Eight potential phosphorylation sites in the cytoplasmic domain.

Six potential N-linked glycosylation sites.

Database accession numbers

	PIR	SWISSPROT	EMBL/GENBANK	REFERENCE
Human αL	S03308	P20701	Y00796	
Human β2	A25967		M15395	
		P05107		18
			Y00057	19
Mouse αL			M60778	17
Mouse β2	S04847	P11835	X14951	

References

[1] **Larson, R.S. and Springer, T.A. (1990) Immunol. Rev. 114, 181–217.**
[2] Hibbs, M.L. et al. (1991) J. Exp. Med. 174, 1227–1238.
[3] Valmu, L. et al. (1991) Eur. J. Immunol. 21, 2857–2862.
[4] Sanchez-Madrid, F. and Corbi, A.L. (1993) Seminars Cell Biol. 3, 199–210.
[5] Dransfield, I. et al. (1992) J. Cell Biol. 116, 219–226.
[6] Van Kooyk, Y. et al. (1991) J. Cell Biol. 112, 345–354.
[7] Dustin, M.L. and Springer, T.A. (1989) Nature 341, 619–624.
[8] Inghirami, G. et al. (1990) Science 250, 682–686.
[9] Anderson, D.C. and Springer, T.A. (1987) Annu. Rev. Med. 38, 175–194.
[10] Marlin, S.D. and Springer, T.A. (1987) Cell 51, 813–819.
[11] Staunton, D.E. et al. (1989) Nature 339, 61–64.
[12] Fougerolles, A.R. and Springer, T.A. (1992) J. Exp. Med. 175, 185–190.
[13] Weitzman, J.B. et al. (1991) FEBS Lett. 294, 97–103.
[14] Corbi, A.L. et al. (1988) J. Exp. Med. 167, 1597–1607.
[15] Marlin, S. et al. (1986) J. Exp. Med. 164, 855–867.
[16] Larson, R.S. et al. (1989) J. Cell Biol 108, 703–712.
[17] Kaufman, Y. et al. (1991) J. Immunol. 147, 369–374.
[18] Kishimoto, T.K. et al. (1987) Cell 48, 681–690.
[19] Law, S.K.A. et al. (1987) EMBO J. 6, 915–919.
[20] van Seventer, G.A. et al. (1990) J. Immunol. 144, 4579–4586.

LFA-3 (leukocyte function associated molecule-3) CD58

Family
Immunoglobulin superfamily. LFA-3 is structurally similar to its ligand CD2, although they share only 21% amino acid identity.

Cellular distribution
Widely expressed including leukocytes, erythrocytes, endothelial cells, epithelial cells and fibroblasts.

Function
LFA-3 mediates the interactions of antigen-presenting cells and target cells, with activated T cells and thymocytes. Binding of LFA-3 to its ligand CD2 enhances T cell stimulation [reviewed in ref.1]. The binding of specific anti-LFA-3 antibodies can induce IL1 release from monocytes, thymic epithelial cells and keratinocytes [2].

Ligand
CD2 [3],

Gene location
1p13.

Structure [4]

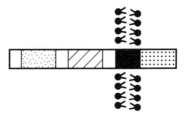

Domain 1 resembles an immunoglobulin V-region but lacks the disulphides characteristic of most Ig-domains.

Molecular weights
Polypeptide   25 339 (transmembrane form); 20 595 (GPI-linked form)
SDS PAGE   55 000–70 000

Amino acid sequence [4]

```
  1  MVAGSDAGRA LGVLSVVCLL HCFGFISCFS QQIYGVVYGN VTFHVPSNVP
 51  LKEVLWKKQK DKVAELENSE FRAFSSFKNR VYLDTVSGSL TIYNLTSSDE
101  DEYEMESPNI TDTMKFFLYV LESLPSPTLT CALTNGSIEV QCMIPEHYNS
151  HRGL1MYSWD CPMEQCKRNS TSIYFKMEND LPQKIQCTLS NPLFNTTSSI
201  ILTTCIPSSG HSRHRYALIP IPLAVITTCI VLYMNGILKC DRKPDRTNSN
```

LFA-3

Domain structure
  1–28   Signal sequence
 41–109   Ig-1
137–191   Ig-2
215–238   Transmembrane domain

In the GPI-linked variant, residues 236 and 237 (GI) are replaced by VL, which forms the COOH-terminus [5]. Ser 208 is the potential attachment site for the GPI-anchor.

Residues 40 and 94 are confirmed glycosylation sites. There are four additional potential sites.

Database accession numbers

	PIR	SWISSPROT	EMBL/GENBANK
Human GPI-linked variant	S01269		X06296
Human trans-membrane variant	A28564	P19256	Y00636

Alternative forms
A transmembrane form of LFA-3 [4] and a form attached to the membrane via a GPI-linkage [5], are produced by alternative splicing.

References
[1] **Dustin, M.L. and Springer, T.A. (1991) Annu. Rev. Immunol. 9, 27–66.**
[2] Le, P.T. et al. (1990) J. Immunol. 144, 4541–4547.
[3] Selveraj, P. et al. (1987) Nature 326, 400–403.
[4] Wallner, B.P. et al. (1987) J. Exp. Med. 166, 923–932.
[5] Seed, B.A. (1987) Nature 329, 840–842.

L-Selectin   Leukocyte adhesion molecule-1 (LAM-1), Leu-8

Other names
TQ1 (human); gp90^{MEL-14}, mLHR (mouse); LECCAM-1, LECAM-1

Family
Selectin (Ca^{2+} dependent lectin).

Cellular distribution
Peripheral blood B and T lymphocytes, neutrophils, immature and some mature thymocytes, monocytes, eosinophils, basophils, bone marrow myeloid progenitor cells and erythroid precursor cells, NK cells [1-3]. Also expressed by some malignant leukocytes [4].

Function
Peripheral lymph node homing receptor. Mediates PMN, lymphocyte and monocyte binding to endothelium at inflammatory sites and lymphocyte binding to HEV of peripheral lymph node. Also participates in the adherence of phagocytes to endothelium [5-8]. Mediates leukocyte rolling in mesenteric venules *in vivo* [9].

Regulation of expression
The regulation of L-selectin has recently been reviewed [10]. It is constitutively expressed on lymphocytes and neutrophils. Activation of leukocytes (by phorbol esters, cytokines or chemoattractants) results in shedding of L-selectin. Shed material can be found at high levels in normal serum [11].

In neutrophils, its downregulation is also stimulated by exposure to IL1 activated endothelial cells. In lymphocytes, expression of L-selectin appears to depend on the activation status and stage of differentiation. Mitogenic stimuli increase L-selectin expression.

Ligands
Unknown. A candidate molecule is MECA-79 antigen (PNad) [12]. The ligand probably requires sialic acid [7,13].

A mouse ligand structure Sgp50 or GLYCAM-1, has recently been cloned [14].

Gene structure

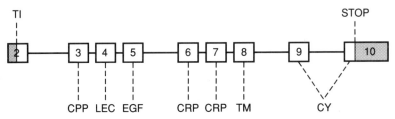

L-Selectin

exon
- 2   5'UT and translation initiation site
- 3   Leader sequence
- 4   Lectin
- 5   EGF
- 6   CR1
- 7   CR2
- 8   transmembrane
- 9   COOH-terminal phosphorylation cassette
- 10 cytoplasmic tail and 3'UT

Spans >30 kb with at least 10 exons [15]. The mouse gene structure is similar [16].

Gene location and size
Human 1q23–25; >30 kb [15].
Mouse chromosome 1 [16]

Structure

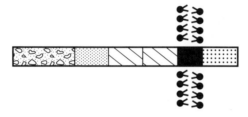

Molecular weights
Polypeptide   37 571
SDS PAGE   74 000 (lymphocyte form) [17]; 90 000–100 000 (neutrophil form) [2]

Amino acid sequence (from human tonsil) [18]

```
                              ↓
  1   MIFPWKCQST QRDLWNIFKL WGWTMLCCDF LAHHGTDCWT YHYSEKPMNW
               *
 51   QRARRFCRDN YTDLVAIQNK AEIEYLEKTL PFSRSYYWIG IRKIGGIWTW
               *
101   VGTNKSLTEE AENWGDGEPN NKKNKEDCVE IYIKRNKDAG KWNDDACHKL
                                     *
151   KAALCYTASC QPWSCSGHGE CVEIINNYTC NCDVGYYGPQ CQFVIQCEPL
                                              *              *
201   EAPELGTMDC THPLGNFSFS SQCAFSCSEG TNLTGIEETT CGPFGNWSSP
                                    *
251   EPTCQVIQCE PLSAPDLGIM NCSHPLASFS FTSACTFICS EGTELIGKKK
                  *
301   TICESSGIWS NPSPICQKLD KSFSMIKEGD YNPLFIPVAV MVTAFSGLAF

351   IIWLARRLKK GKKSKRSMND PY
```

L-Selectin

Domain structure
    1–28   Signal sequence
    29–38   Propeptide
    39–156   Lectin domain
    160–192 EGF domain
    193–255 CRP repeat
    256–316 CRP repeat
    333–355 Transmembrane domain
    356–372 Cytoplasmic domain

Seven potential N-linked glycosylation sites (human).
Two cytoplasmic Ser potential phosphorylation sites (human) for protein kinase C.

Database accession numbers
	PIR	SWISSPROT	EMBL/GENBANK	REFERENCE
Human	A33912			
		P14151	X16150	18
			M25280	20
			X17519	21
Mouse	A32375			
		P18337	M36005	19

Alternative forms
None known.

References
1 Tedder, T.F. et al. (1990) J. Immunol. 144, 532–540.
2 Griffin, J.D. et al. (1990) J. Immunol. 145, 576–584.
3 Gallatin, W.M. et al. (1983) Nature 304, 30–34.
4 Spertini, O. et al. (1991) Leukemia 5, 300–308.
5 **Geoffroy, J.S. and Rosen, S.D. (1989) J. Cell Biol. 109, 2463–2469.**
6 Lasky, A. et al. (1989) Cell 56, 1045–1055.
7 Spertini, O. et al. (1991) J. Immunol. 147, 2565–2573.
8 Spertini, O. et al. (1992) J. Exp. Med. 175, 1789–1792.
9 Ley, K. et al. (1991) Blood 77, 2553–2555.
10 **Kansas, G.S. et al. (1991) In Cellular and Molecular Mechanisms of Inflammation, Vol. 2, Academic Press, New York, pp.31–59.**
11 Schleiffenbaum, B. et al. (1993) J. Cell Biol. (in press).
12 Berg, E.L. et al. (1991) J. Cell Biol. 114, 343–349.
13 True, D.D. et al. (1990) J. Cell Biol. 111, 2725–2764.
14 Lasky, L.A. et al. (1992) Cell 69, 927–938.
15 Ord, D.C. et al. (1990) J. Biol. Chem. 265, 7760–7767.
16 Dowbenko, D. J. et al. (1991) Genomics 9, 270–277.
17 Tedder, T.F. et al. (1990) Eur. J. Immunol. 20, 1357–1366.
18 Tedder, T.F. et al. (1989) J. Exp. Med. 170, 123–133.
19 Siegelman, M.H. et al. (1989) Science 243, 1165–1172.
20 Siegelman, M.H. and Weissman, I.L. (1989) Proc. Natl Acad. Sci. USA 86, 5562–5566.
21 Camerini, S.D. et al. (1989) Nature 342, 78–82.

M-Cadherin

Muscle cadherin

Family
Cadherin.

Cellular distribution
Differentiating mouse muscle cells; low level of expression in myoblasts [1].

Function
Unknown, but based on the primary amino acid sequence it is most likely to act as a Ca2+ dependent cell adhesion molecule [1].

Regulation of expression
Upregulated after induction of myotube formation [1].

Ligand
Unknown. Does not contain the N-terminal HAV sequence which is conserved in other cadherins [1].

Gene location
Mouse chromosome 8 (unpublished data).

Gene structure
Unknown.

Molecular weights
Polypeptide   89 089
SDS PAGE   104 000 (predicted from amino acid sequence)

Amino acid sequence (from mouse C2 myoblasts and C2 myotubes) [1]

```
  1   KDSAQCSQRG GRRSTTARST GRPVALVLGS SATASMGSAL LLPRLLAQSL

 51   GLSWAVPEPE PSTLYPWMPA SAPGRVRRAW VIPPISVSEN HKRLPYPLVQ
                                                      *
101   IKSDKQQLGS VIYSIQGPGV DEEPRNVFSI DKFTGRVYLN ATLDREKTDR

151   FRLRAFALDL GGSTLEDPTD LEIVVVDQND NRPAFLQDVF RGHILEGAIP

201   GTFVTRAEAT DADPETDNA ALRFSILEQG SPEFFSIDEH TGEIRTVQVG
                  *
251   LDREVVAVYN LTLQVADMSG DGLTATASAI ISIDDINDNA PEFTKDEFFM

301   EAAEAVSGVD VGRLEVEDKD LPGSPNWVAR FTILEGDPDG QFKIYTDPKT
```

M-Cadherin

```
351  NEGVLSVVKP LDYESREQYE LRVSVQNEAP LQAAAPRARR GQTRVSVWVQ
401  DTNEAPVFPE NPLRTSIAEG APPGTSVATF SARDPDTEQL QRISYSKDYD
451  PEDWLQVDGA TGRIQTQRVL SPASPFLKDG WYRAIILALD NAIPPSTATG
501  TLSIEILEVN DHAPALALPP SGSLCSEPDQ GPGLLLGATD EDLPPHGAPF
                           *          *
551  HFQLNPRVPD LGRNWSVSQI NVSHARLRLR HQVSEGLHRL SLLLQDSGEP
                *
601  PQQREQTLNV TVCRCGSDGT CLPGAAALRG GGVGVSLGAL VIVLASTVVL
651  LVLILFAALR TRFRGHSRGK SLLHGLQEDL RDNILNYDEQ GGGEEDQDAY
701  DINQLRHPVE PRATSRSLGR PPLRRDAPFS YVPQPHRVLP TSPSDIANFI
751  SDGLEAADSD PSVPPYDTAL IYDYEGDGSV AGTLSSILSS LGDEDQDYDY
801  LRDWGPRFAR LADMYGHQ
```

Domain structure

89–185	EC1
186–293	EC2
294–408	EC3
409–514	EC4
515–623	EC5
624–661	Transmembrane domain
662–818	Cytoplasmic domain

Five potential N-linked glycosylation sites in the extracellular domain.

Three Ca^{2+} binding sites in the extracellular domain (residues 177–181, 211–213 and 285–289).

Database accession numbers

	PIR	SWISSPROT	EMBL/GENBANK
Mouse			M74541

Alternative forms

None known.

Reference

[1] Donalies, M. et al. (1991) Proc. Natl Acad. Sci. USA 88, 8024–8028

Mac-1 CR3, CD11b/CD18, αMβ2

Other names
Leukocyte adhesion receptor Mo1, neutrophil adherence receptor

Family
β2 integrin (dimer of αM and β2 subunits).

Cellular distribution
Blood monocytes, macrophages and granulocytes. NK cells, PMA-induced myeloid cell lines (not B and T cell lineages).

Function
The functions of Mac-1 have recently been reviewed in ref. 1.

Mac-1 mediates PMN and monocyte adherence to endothelium and subsequently PMN extravasation to sites of inflammation [2,3]. It also mediates phorbol ester-induced PMN homotypic adhesion and chemotaxis [4], and is involved in phagocytosis of complement-coated particles bound via CR1 and IgGFcR [5]. A Ca^{2+}-dependent conformational change allows distinction of the roles of Mac-1 in phagocytosis and adhesion [6].

Recent studies have indicated that a lipid, integrin modulating factor (IMF-1) can enhance binding of Mac-1 to its ligand by effecting a conformational change to a high avidity state [7].

Regulation of expression
Differentiation and maturation of monocytes and myelomonocytic cell lines leads to increased surface expression of Mac-1. It is also upregulated on neutrophils and monocytes by inflammatory stimuli. Mac-1 is stored in intracellular vesicles which are rapidly mobilized to the cell surface in response to chemoattractants such as fMLP, C5a and leukotriene B4 [8].

Ligands
C3bi, ICAM-1 (via its third Ig-domain) [9,10], factor X [11], fibrinogen [12].

Gene structure
The promoter region has been characterized. The CD11b gene contains a single transcription initiation site. The region 242 bp upstream and 71 bp downstream of this site are sufficient to direct tissue-, cell- and development-specific expression *in vitro*, which mimics that of CD11b *in vivo* [13]. The 5' flanking region, upstream of the initiation site, also contains AP-1, AP-2 and AP-3-like binding sites, 4 *Alu* repeats and an $(AC)_{16}$ sequence. The latter may have played a role in the gene duplication process by which the CD11b gene is thought to have arisen.

Gene location
16p11–13.1 (αM) [14].
21q22 (β2).

Mac-1

Structure

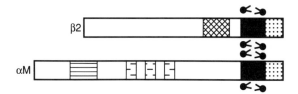

Molecular weights
Polypeptide   125 611 (αM)
SDS PAGE   170 000 (αM)

Amino acid sequence (αM) (from PMA-induced HL-60 cells) [15–17]

```
  1  MALRVLLLTA LTLCHGFNLD TENAMTFQEN ARGFGQSVVQ LQGSRVVVGA
 51  PQEIVAANQR GSLYQCDYST GSCEPIRLQV PVEAVNMSLG LSLAATTSPP
101  QLLACGPTVH QTCSENTYVK GLCFLFGSNL RQQPQKFPEA LRGCPQEDSD
151  IAFLIDGSGS IIPHDFRRMK EFVSTVMEQL KKSKTLFSLM QYSEEFRIHF
201  TFKEFQNNPN PRSLVKPITQ LLGRTHTATG IRKVVRELFN ITNGARKNAF
251  KILVVITDGE KFGDPLGYED VIPEADREGV IRYVIGVGDA FRSEKSRQEL
301  NTIASKPPRD HVFQVNNFEA LKTIQNQLRE KIFAIEGTQT GSSSSFEHEM
351  SQEGFSAAIT SNGPLLSTVG SYDWAGGVFL YTSKEKSTFI NMTRVDSDMN
401  DAYLGYAAAI ILRNRVQSLV LGAPRYQHIG LVAMFRQNTG MWESNANVKG
451  TQIGAYFGAS LCSVDVDSNG STDLVLIGAP HYYEQTRGGQ VSVCPLPRGQ
501  RARWQCDAVL YGEQGQPWGR FGAALTVLGD VNGDKLTDVA IGAPGEEDNR
551  GAVYLFHGTS GSGISPSHSQ RIAGSKLSPR LQYFGQSLSG GQDLTMDGLV
601  DLTVGAQGHV LLLRSQPVLR VKAIMEFNPR EVARNVFECN DQVVKGKEAG
651  EVRVCLHVQK STRDRLREGQ IQSVVTYDLA LDSGRPHSRA VFNETKNSTR
701  RQTQVLGLTQ TCETLKLQLP NCIEDPVSPI VLRLNFSLVG TPLSAFGNLR
751  PVLAEDAQRL FTALFPFEKN CGNDNICQDD LSITFSFMSL DCLVVGGPRE
```

```
 801  FNVTVTVRND GEDSYRTQVT FFFPLDLSYR KVSTLQNQRS QRSWRLACES
                                  *
 851  ASSTEVSGAL KSTSCSINHP IFPENSEVTF NITFDVDSKA SLGNKLLLKA
         *          *                             *          *
 901  NVTSENNMPR TNKTEFQLEL PVKYAVYMVV TSHGVSTKYL NFTASENTSR
                                  *                    *
 951  VMQHQYQVSN LGQRSLPISL VFLVPVRLNQ TVIWDRPQVT FSENLSSTCH
                                  *                    *
1001  TKERLPSHSD FLAELRKAPV VNCSIAVCQR IQCDIPFFGI QEEFNATLKG
         *                        *
1051  NLSFDWYIKT SHNHLLIVST AEILFNDSVF TLLPGQGAFV RSQTETKVEP
1101  FEVPNPLPLI VGSSVGGLLL LALITAALYK LGFFKRQYKD MMSEGGPPGA
1151  EPQ
```

Domain structure

αM contains seven internal homologous repeats with an additional I-domain found between repeats II and III. Three divalent metal cation binding sites. The αM chain is constitutively phosphorylated.

1–16	Signal sequence
17–1105	Extracellular domain
20–74	Repeat I
75–127	Repeat II
342–389	Repeat III
390–442	Repeat IV
443–507	Repeat V
508–569	Repeat VI
570–643	Repeat VII
164–350	I-domain
1106–1129	Transmembrane domain
1130–1153	Cytoplasmic domain

Nineteen potential N-linked glycosylation sites.

Amino acid sequence (β2)

See LFA-1 entry.

Database accession numbers

	PIR	SWISSPROT	EMBL/GENBANK	REFERENCE
Human αM	A31108	P11215		
			J03925	2
			M18044	1
			X07421	1
Mouse αM	S00551	P05555	X07640	
Guinea-pig αM	B30892	P11578	M19663	

References

1 **Larson, R.S. and Springer, T.A. (1990) Immunol. Rev. 114, 181–217.**
2 Harlan, J.M. et al. (1985) Blood 66, 167–178.
3 Arnaout, M.A. et al. (1988) J. Cell Physiol. 137, 305.
4 Anderson, D.C. et al. (1986) J. Immunol. 137, 15–27.
5 Carlos, T.M. and Harlan, J.M. (1990) Immunol. Rev. 114, 1–24.
6 Graham, I.L. and Brown, E.J. (1991) J. Immunol. 146, 685–691.
7 Hermanowski-Vosatka, A. et al. (1992) Cell 68, 341–352.
8 Miller, L.J. et al. (1987) J. Clin. Invest. 80, 535–544.
9 **Sanchez-Madrid, F. and Corbi, A.L. (1993) Seminars Cell Biol. 3, 199–210.**
10 Diamond, M.S. et al. (1991) Cell 65, 961–971.
11 Altieri, D.C. et al. (1988) Proc. Natl Acad. Sci. USA 85, 7462–7466.
12 Wright, S.D. et al. (1988) Proc. Natl Acad. Sci. USA 7734–7738.
13 Shelley, C.S. and Arnaout, M.A. (1991) Proc. Natl Acad. Sci. USA 88, 10525–10529.
14 Corbi, A.L. et al. (1988) J. Exp. Med. 167, 1597–1607.
15 Arnaout, M.A. et al. (1988) J. Cell Biol. 106, 2153–2158.
16 Corbi, A.L. et al. (1988) J. Biol. Chem. 263, 12403–12411.
17 Hickstein, D.D. et al. (1989) Proc. Natl Acad. Sci. USA 86, 257–261.

Myelin-associated glycoprotein (MAG)

Family
Immunoglobulin superfamily, C2 subset.

Cellular distribution
Oligodendrocytes, Schwann cells.

Function
The pattern of expression and cellular distribution suggests a role for MAG in the formation and maintenance of the myelin sheath [reviewed in ref. 1]. Antibodies to MAG disrupt neuron–oligodendrocyte and oligodendrocyte–oligodendrocyte interactions *in vitro* [2]. MAG incorporated into liposomes binds to cultured neurons [3].

Regulation of expression
The expression of L- and S-MAG is developmentally regulated. In the CNS, L-MAG is the predominant isoform found in early development, whereas S-MAG is the major isoform found in the adult. In the PNS L-MAG is moderately expressed early in development, but is virtually absent in the adult [4].

Ligands
Collagens I–VI, IX and G, heparin [5]. The receptor for MAG on axons has not been identified.

Gene location and size
19q12–19q13.2 (human), 16 kb

Gene structure[6]

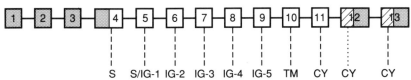

Exon 12 contains a stop codon and is present in S-MAG, but not L-MAG, leading to the expression of a truncated cytoplasmic domain. Exon 2 is not present in S-MAG.

Structure[1]

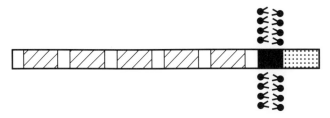

Myelin-associated glycoprotein

Molecular weights
Polypeptide   67 173 (L-MAG); 62 384 (S-MAG)
SDS Page   L-MAG and S-MAG migrate as a band of M_r 100 000

Amino acid sequence (L-MAG) [6]

```
                        ↓
  1  MIFLTTLPLF WIMISASRGG HWGAWMPSSI SAFEGTCVSI PCRFDFPDEL
                                                          *
 51  RPAVVHGVWY FNSPYPKNYP PVVFKSRTQV VHESFQGRSR LLGDLGLRNC

101  TLLISTLSPE LGGKYYFRGD LGGYNQYTFS EHSVLDIINT PNIVVPPEVV

151  AGTEVEVSCM VPDNCPELRP ELSWLGHEGL GEPTVLGRLR EDEGTWVQVS
                                 *                      *
201  LLHFVPTREA NGHRLGCQAA FPNTTLQFEG YASLDVKYPP VIVEMNSSVE

251  AIEGSHVSLL CGADSNPPPL LTWMRDGMVL REAVAESLYL DLEEVTPAED
                    *                *
301  GIYACLAENA YGQDNRTVEL SVMYAPWKPT VNGTVVAVEG ETVSILCSTQ

351  SNPDPILTIF KEKQILATVI YESQLQLELP AVTPEDDGEY WCVAENQYGQ
            *                                            *
401  RATAFNLSVE FAPIILLESH CAAARDTVQC LCVVKSNPEP SVAFELPSRN
        *
451  VTVNETEREF VYSERSGLLL TSILTLRGQA QAPPRVICTS RNLYGTQSLE

501  LPFQGAHRLM WAKIGPVGAV VAFAILIAIV CYITQTRRKK NVTESPSFSA
                      ‾‾‾‾‾‾‾‾‾‾‾‾‾‾‾‾‾‾‾‾
551  GDNPHVLYSP EFRISGAPDK YESEKRLGSE RRLLGLRGEP PELDLSYSHS

601  DLGKRPTKDS YTLTEELAEY AEIRVK
```

Domain structure
    1–19   Signal sequence
   35–104   Ig-1
  152–221   Ig-2
  254–309   Ig-3
  340–396   Ig-4
  425–492   Ig-5
  517–536   Transmembrane domain
Eight potential N-linked glycosylation sites.

Amino acid sequence (S-MAG) [6]
S-MAG is identical to L-MAG in the extracellular and transmembrane domains. The cytoplasmic domain differs after residue 573, where the N-terminal residues are: REVSTRDCH.

Database accession numbers

	PIR	SWISSPROT	EMBL/GENBANK	REFERENCE
Rat L-MAG		P07722		
Rat S-MAG	B29028	P07723	M14871	6
Mouse L-MAG	B33785	P20917	M16800	6
Mouse S-MAG	A33785	P16880		
Human	A33263	P20916	M31811	7
			M29273	8

Alternative forms

Alternative splicing generates two variants, S-MAG and L-MAG, which differ in the cytoplasmic domain [9].

References

1 **Salzer, J.L. et al. (1990) Ann. New York Acad. Sci. 605, 303–312.**
2 Poltorak, M. et al. (1987) J. Cell Biol. 105, 1893–1899.
3 Sadoul, R. et al. (1990) J. Neurosci. Res. 25, 1–13.
4 Pedraza, C. et al. (1991) J. Neurosci. Res. 29, 141–148.
5 Fahrig, T. et al. (1987) EMBO J. 6, 2875–2883.
6 Lai, C. et al. (1987) Proc. Natl Acad. Sci. USA 84, 4337–4341.
7 Fujita, N. et al. (1989) Biochem. Biophys. Res Commun 165, 1162–1169.
8 Sato, S. et al. (1989) Biochem. Biophys. Res Commun 163, 1473–1480.
9 Salzer, J.L. et al. (1987) J. Cell Biol. 104, 957–965.

N-Cadherin
Neural cadherin, A-CAM, N-Cal-CAM

Family
Cadherin.

Cellular distribution
Neural tissue, lens, cardiac and skeletal muscle.

Function
Ca^{2+} dependent neural cell adhesion protein. May play a role in cell sorting processes. N-Cadherin may participate in the selective attachment of neurites to particular cell types and in guiding the growth cone to specific targets [1,2]. N-Acetyl galactosaminyl transferase (NAC Gal PTase) associates with N-Cadherin in the neural retina and is thought to modulate N-cadherin function [3]. N-Cadherin has also been shown to associate with α, β and γ catenins in chicken cells [4].

Regulation of expression
Developmentally regulated. N-Cadherin appears in neural plate during its invagination and the appearance is coordinated with the disappearance of E-cadherin. After the formation of the neural tube, N-cadherin becomes the major cadherin. There are regional variations in N-cadherin expression during differentiation [5].

Ligands
N-Cadherin (homotypic adhesion). Amino acid residues flanking the HAV sequence located within the N-terminal appear to confer binding specificity.

Gene location
18 [6].

Gene structure
Unknown.

Molecular weights
Polypeptide   82 116
SDS PAGE   130 000

N-Cadherin

Amino acid sequence (from Kelly (human) neuroblastoma cell line)[1]

```
              ↓
  1  MCRIAGALRT LLPLLLALLQ ASVEASGEIA LCKTGFPEDV YSAVLSKDVH

 51  EGQPLLNVKF SNCNGKRKVQ YESSEPADFK VDEDGMVYAV RSFPLSSEHA

101  KFLIYAQDKE TQEKWQVAVK LSLKPTLTEE SVKESAEVEE IVFPRQFSKH
                                                    *
151  SGHLQRQKRD WVIPPINLPE NSRGPFPQEL VRIRSDRDKN LSLRYTVTGP

201  GADQPPTGIF IINPISGQLS VTKPLDREQI ARFHLRAHAV DINGNQVENP
                                *
251  IDIVINVIDM NDNRPEFLHQ VWNGTVPEGS KPGTYVMTVT AIDADDPNAL
                                *
301  NGMLRYRIVS QAPSTPSPNM FTINNETGDI ITVAAGLDRE NVQQYTLIIQ

351  ATDMEGIPTY GLSNTATAVI TVTDVNDNPP EFTAMTFYGE VPENRVDIIV
       *
401  ANLTVTDKDQ PHTPAWNAVY RISGGDPTGR FAIQTDPNSN DGLVTVVKPI

451  DFETNRMFVL TVAAENQVPL AKGIQHPPQS TATVSVTVID VNENPYFAPN

501  PKIIRQEEGL HAGTMLTTFT AQDPDRYMQQ NIRYTKLSDP ANWLKIDPVN
                                         *
551  GQITTIAVLD RESPNVKNNI YNATFLASDN GIPPMSGTGT LQIYLLDIND

601  NAPQVLPQEA ETCETPDPNS INITALDYDI DPNAGPFAFD LPLSPVTIKR
       *                                              *
651  NWTITRLNGD FAQLNLKIKF LEAGIYEVPI IITDSGNPPK SNISILRVRV

701  CQCDFNGDCT DVDRIVGAGL GTGAIIAILL CIIILLILVL MFVVWMKRRD

751  KERQAKQLLI DPEDDVRDNI LKYDEEGGGE EDQDYDLSQL QQPDTVEPDA

801  IKPVGIRRMD ERPIHAEPQY PVRSAAPHPG DIGDFINEGL KAADNDPTAP

851  PYDSLLVFDY EGSGSTAGSL SSLNSSSSGG EQDYDYLNDW GPRFKKLADM

901  YGGGDD
```

Domain structure

1–23	Signal peptide
24–159	Propeptide
160–724	Extracellular domain
170–267	Repeat I
268–382	Repeat II

113

N-Cadherin

383–497   Repeat III
498–603   Repeat IV
604–714   Repeat V
725–746   Transmembrane domain
747–906   Cytoplasmic domain
Seven potential N-linked glycosylation sites.
Three Ca^{2+} binding sites in extracellular domain (residues 259–263, 293–295 and 374–378).
Consensus sequence for precursor cleavage (156–159) [8].

Database accession numbers

	PIR	SWISSPROT	EMBL/GENBANK	REFERENCE
Human		P19022		7
			X54315	
Mouse	A32759		M31131	
		P15116		9
Bovine		P19534	X53615	
Chicken	A29964	P10288	X07277	
X. laevis		P20310		

Alternative forms

Unknown

References
1 Hatta, K. and Takeichi, M. (1986) Nature 320, 447–449.
2 Bixby, J.L. et al. (1987) Proc. Natl Acad. Sci. USA 84, 2555–2559.
3 **Ranscht, B. (1991) Seminars Neurosci. 3, 285–296.**
4 Ozawa, M. and Kemler, R. (1992) J. Cell Biol. 116, 989–996.
5 Takeichi, M. (1988) Development 102, 639–655.
6 Walsh, F.S. et al. (1990) J. Neurochem. 805–812.
7 Reid, R.A. and Hemperley, J.J. (1990) Nucleic Acids Res. 18, 5896.
8 Ozawa, M. and Kemler, R. (1990) J. Cell Biol. 111, 1645–1650.
9 **Miyatani, S. et al. (1989) Science 245, 631–635.**

NCAM (neural cell adhesion molecule) D2-CAM, CD56, Leu-19, NKH1

Family
Immunoglobulin superfamily, C2 subset.

Cellular distribution
Neurons, astrocytes, Schwann cells, myoblasts, myotubes, NK cells, subset of activated T lymphocytes.

Function
NCAM has been shown to be involved in a diverse range of contact mediated interactions between neurons, neurons and astrocytes, neurons and oligodendrocytes and neuronal processes and myotubes [reviewed in ref.1]. NCAM is widely but transiently expressed early in embryogenesis and is assumed to have an important role in the development of normal tissue architecture [reviewed in ref.2]. NCAM has been reported to have a role in contact-dependent inhibition of cell growth [3]. There is evidence that the mechanism by which NCAM promotes neurite outgrowth is through the activation of a G-protein which may open calcium channels [4].

Regulation of expression
NCAM is transiently expressed in many tissues during development [2]. There are also developmentally regulated switches in the expression of isoforms.

Ligands
NCAM binding is homophilic. Also binds heparan sulphate and heparin [5]. NCAM may interact in the plane of the membrane with L1 [6].

Gene structure [1]

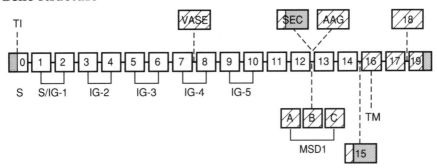

The exon labelled SEC contains a stop codon and its expression leads to the production of a secreted variant. The AAG exon encodes a single amino acid. Expression of exon 15 leads to the production the GPI-anchored variant. Exon 18 encodes the cytoplasmic domain specific to NCAM-180. The VASE and MSD1 exons are described below.

Gene location and size [1]
11q23 (human), 9 (mouse); >100 kb.

Structure (NCAM-180)

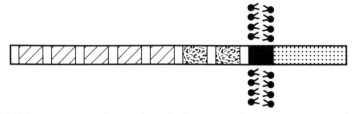

NCAM-140 has a truncated cytoplasmic domain and NCAM-120 is anchored in the membrane via a GPI-linkage.

Molecular weights
Polypeptide   117 570 (NCAM-180), 90 998 (NCAM-140), 79 567 (NCAM-120)
SDS PAGE   180 000, 140 000, 120 000

Amino acid sequence (NCAM-120)
NCAM-120 is attached to the membrane via a GPI-linkage [12]

```
  1  MLRTKDLIWT LFFLGTAVSL QVDIVPSQGE ISVGESKFFL CQVAGDAKDK
 51  DISWFSPNGE KLSPNQQRIS VVWNDDDSST LTIYNANIDD AGIYKCVVTA
101  EDGTQSEATV NVKIFQKLMF KNAPTPQEFK EGEDAVIVCD VVSSLPPTII
151  WKHKGRDVIL KKDVRFIVLS NNYLQIRGIK KTDEGTYRCE GRILARGEIN
201  FKDIQVIVNV PPTVQARQSI VNATANLGQS VTLVCDADGF PEPTMSWTKD
251  GEPIENEEED ERSRSSVSDS SEVTIRNVDK NDEAEYVCIA ENKAGEQDAS
301  IHLKVFAKPK ITYVENQTAM ELEEQVTLTC EASGDPIPSI TWRTSTRNIS
351  SEEQDLDGHM VVRSHARVSS LTLKSIQYRD AGEYMCTASN TIGQDSQSID
401  LEFQYAPKLQ GPVAVYTWEG NQVNITCEVF AYPSATISWF RDGQLLPSSN
451  YSNIKIYNTP SASYLEVTPD SENDFGNYNC TAVNRIGQES LEFILVQADT
501  PSSPSIDRVE PYSSTAQVQF DEPEATGGVP ILKYKAEWKS LGEESWHFTW
551  YDAKEANMEG IVTIMGLKPE TTYSDRLAAL NGKGLGEIMQ PSESKTQPVP
601  ELSAPKLEGQ MGEDGNSIKV NLIKQDDGGS PIRHYLVKYR ALASEWKPEI
651  RLPSGSHHVM LKSLDWNAEY EVYVVAENQQ GKSKAAHFVF RTSAQPTAIP
701  ATLGGSSTSY TLVSLLFSAV TLLLL
```

NCAM

Domain structure

1–19	Signal sequence
34–100	Ig-1
132–193	Ig-2
228–292	Ig-3
323–390	Ig-4
420–484	Ig-5
520–605	FN-1
620–700	FN-2

Ser 707 is the potential attachment site for the GPI anchor.

Amino acid sequence (NCAM-140)

The C-terminal region of NCAM-140, which has a transmembrane sequence and a short cytoplasmic domain [13].

```
  1   LASEWKPEIR LPSGSDHVML KSLDWNAEYE VYVVAENQQG KSKAAHFVFR

 51   TSAQPTAIPA NGSPTAGLST GAIVGILIVI FVLLLVVMDI TCYFLNKCGL

101   LMCIAVNLCG KAGPGAKGKD MEEGKAAFSK DESKEPIVEV RTEEERTPNH

151   DGGKHTEPNE TTPLTEPEKG PVETKSEPPE SEAKPAPTEV KTVPNDATQT

201   KENESKA
```

Domain structure

67–87   Transmembrane domain

Amino acid sequence (NCAM-180)

The C-terminal region of NCAM-180, which contains an additional exon in the cytoplasmic domain [13].

```
  1   LASEWKPEIR LPSGSDHVML KSLDWNAEYE VYVVAENQQG KSKAAHFVFR

 51   TSAQPTAIPA NGSPTAGLST GAIVGILIVI FVLLLVVMDI TCYFLNKCGL

101   LMCIAVNLCG KAGPGAKGKD MEEGKAAFSK DESKEPIVEV RTEEERTPNH

151   DGGKHTEPNE TTPLTEPELP ADTTATVEDM LPSVTTVTTN SDTITETFAT

201   AQNSPTSETT TLTSSIAPPA TTVPDSNSVP AGQATPSKGV TASSSSPASA

251   PKVAPLVDLS DTPTSAPSAS NLSSTVLANQ GAVLSPSTPA SAGETSKAPP

301   ASKASPAPTP TPAGAASPLA AVAAPATDAP QAKQEAPSTK GPDPEPTQPG
```

NCAM

```
351  TVKNPPEAAT APASPKSKAA TTNPSQGEDL KMDEGNFKTP DIDLAKDVFA
401  ALGSPRPATG ASGQASELAP SPADSAVPPA PAKTEKGPVE TKSEPPESEA
451  KPAPTEVKTV PNDATQTKEN ESKA
```

Domain structure

68–87   Transmembrane domain

Database accession numbers

	PIR	SWISSPROT	EMBL/GENBANK	REFERENCE
Human NCAM-125 (GPI-linked)	A26883	P13592	X16841	14
Human NCAM-140	B26883	P13591	M17410	15
Human NCAM (soluble)	A31635	P13593	M22094	16
Mouse NCAM-120 (GPI-linked)	A29673	P13594	Y00051	12
Mouse NCAM-140	S00844	P13595	X06328	13
Mouse NCAM-180	S00384		X071987	13
Rat	S00846	P13596	X06564	17
Chicken	A25435	P13590	M13210	18

NCAM-125 is attached to the membrane via a GPI-link and is homologous to NCAM-120 [14].

Alternative forms

Four major isoforms of NCAM have been identified, which are generated by alternative splicing of a single gene. Two of these isoforms are transmembrane proteins, differing only in their cytoplasmic domains. NCAM-140 has a short cytoplasmic tail, whereas NCAM-180 has a large cytoplasmic domain thought to interact with the cytoskeleton. NCAM-180 expression is restricted to neural tissues. The third major isoform, NCAM-120, lacks a transmembrane domain, but is attached to the membrane via a glycosylphosphotidylinositol (GPI)-linkage [reviewed in ref.1]. One role of the different isoforms may be to target them to different cellular destinations [7], or to alter their capacity to activate intracellular signalling pathways. A fourth variant has been identified in which the transmembrane domain is deleted, leading to the expression of a soluble molecule [8]. Alternative splicing of additional exons generates many further variants. Important additional exons include a group of three exons collectively called MSD1 [9] and the VASE exon. Expression of the VASE exon converts the fourth Ig-domain from a C2- to V-type [10] and downregulates the neurite growth-promoting activity of NCAM-140 [11].

Further variants are produced by post-translational modifications including major changes in the degree of sialylation and variations in the pattern of phosphorylation and sulphation [1,2].

References
1. **Walsh, S.F. and Docherty, P. (1992) Seminars Neurosci. 3, 271–284.**
2. **Edelman, G.M. (1986) Annu. Rev. Cell Biol. 2, 81–116.**
3. Aoki, J. et al. (1991) J. Cell Biol. 115, 1751–1761.
4. Docherty, P. et al. (1991) Cell 67, 21–33.
4. Nitta, T. et al. (1989) J. Exp. Med. 170, 1759.
5. Cole, G.J. and Akeson, R. (1989) Neuron 2, 1157–1165.
6. Kadmon, G. et al. (1990) J. Cell Biol. 110, 209–218.
7. Powell, S.K. et al. (1991) Nature 353, 76–77.
8. Gower, H. et al. (1988) Cell 55, 955–964.
9. Thompson, J. et al. (1989) Genes Dev. 3, 348–357.
10. Akeson, R. (1990) J. Cell Biol. 115, 2089–2096.
11. Doherty, P. et al. (1992) Nature 356, 791–793.
12. Barthels, R. et al. (1987) EMBO J. 6, 907–914.
13. Barbas, J.A. et al. (1988) EMBO J. 7, 625–632.
14. Barton, C.H. et al. (1990) Development 104, 165–173.
15. Dickson, G. et al. (1990) Cell 50, 1119–1130.
16. Gower, H.J. et al. Cell 55, 955–964.
17. Small, S.J. et al. (1987) J. Cell Biol. 111, 2335–2345.
18. Cunningham, B.A. et al. (1987) Science 236, 799–806.

NEUROGLIAN

Family
Immunoglobulin superfamily, C2 subset. Neuroglian shows homology to mouse L1 [reviewed in refs 1, 2].

Cellular distribution
Two variants of neuroglian show different distributions. A form with a short cytoplasmic tail is widely expressed. A form with a long cytoplasmic domain is restricted to neurons in the CNS and neurons and some glial cells in the PNS [3].

Function
The homology with mouse L1 suggests neuroglian mediates the adhesion of neurons and glia within the nervous system. The short form may have a more general role in adhesion.

Ligand
Neuroglian functions as a homophilic adhesion molecule when expressed in S2 cells. Adhesion is calcium-independent.

Gene location [4]
7F.

Structure [4]

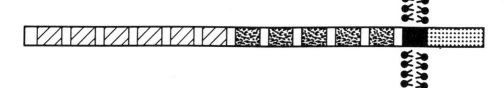

Molecular weights
Polypeptide   135 897
SDS PAGE   180 000, 167 000

Amino acid sequence (from 12 hour embryo, Long form) [4]

```
  1   MWRQSTILAA LLVALLCAGS AESKGNRPPR ITKQPAPGEL LFKVAQQNKE

 51   SDNPFIIECE ADGQPEPEYS WIKNGKKFDW QAYDNRMLRQ PGRGTLVITI

101   PKDEDRGHYQ CFASNEFGTA TSNSVYVRKA ELNAFKDEAA KTLEAVEGEP

151   FMLKCAAPDG FPSPTVNWMI QESIDGSIKS INNSRMTLDP EGNLWFSNVT
```

Neuroglian

```
 201  REDASSDFYY ACSATSVFRS EYKIGNKVLL DVKQMGVSAS QNKHPPVRQY
 251  VSRRQSLALR GKRMELFCIY GGTPLPQTVW SKDGQRIQWS DRITQGHYGK
 301  SLVIRQTNFD DAGTYTCDVS NGVGNAQSFS IILNVNSVPY FTKEPEIATA
 351  AEDEEVVFEC RAAGVPEPKI SWIHNGKPIE QSTPNPRRTV TDNTIRIINL
                      *                                    *
 401  VKGDTGNYGC NATNSLGYVY KDVYLNVQAE PPTISEAPAA VSTVDGRNVT
 451  IKCRVNGSPK PLVKWLRASN WLTGGRYNVQ ANGDLEIQDV TFSDAGKYTC
 501  YAQNKFGEIQ ADGSLVVKEH TRITQEPQNY EVAAGQSATF RCNEAHDDTL
 551  EIEIDWWKDG QSIDFEAQPR FVKTNDNSLT IAKTMELDSG EYTCVARTRL
 601  DEATARANLI VQDVPNAPKL TGITCQADKA EIHWEQQGDN RSPILHYTIQ
           *                              *
 651  FNTSFTPASW DAAYEKVPNT DSSFVVQMSP WANYTFRVIA FNKIGASPPS
 701  AHSDSCTTQP DVPFKNPDNV VGQGTEPNNL VISWTPMPEI EHNAPNFHYY
 751  VSWKRDIPAA AWENNNIFDW RQNNIVIADQ PTFVKYLIKV VAINDRGESN
                                *
 801  VAAEEVVGYS GEDRPLDAPT NFTMRQITSS TSGYMAWTPV SEESVRGHFK
 851  GYKIQTWTEN EGEEGLREIH VKGDTHNALV TQFKPDSKNY ARILAYNGRF
 901  NGPPSAVIDF DTPEGVPSPV QGLDAYPLGS SAFMLHWKKP LYPNGKLTGY
 951  KIYYEEVKES YVGERREYDP HITDPRVTRM KMAGLKPNSK YRISITATTK
1001  MGEGSEHYIE KTTLKDAVNV APATPSFSWE QLPSDNGLAK FRINWLPSTE
1051  GHPGTHFFTM HRIKGETQWI RENEEKNSDY QEVGGLDPET AYEFRVVSVD
                                                      *
1101  GHFNTESATQ EIDTNTVEGP IMVANETVAN AGWFIGMMLA LAFIIILFII
1151  ICIIRRNRGG KYDVHDRELA NGRRDYPEEG GFHEYSQPLD NKSAGRQSVS
1201  SANKPGVESD TDSMAEYGDG DTGQFTEDGS FIGQYVPGKL QPPVSPPQPL
1251  NNSAAAHQAA PTAGGSGAAG SAAAAGASGC ASSAGGAAAS NGGAAAGAVA
1301  TYV
```

Domain structure

1–23	Signal sequence
52–115	Ig-1
148–216	Ig-2
261–321	Ig-3
353–414	Ig-4
446–504	Ig-5
535–598	Ig-6
634–717	FN-III
734–819	FN-III
837–917	FN-III
937–1004	FN-III
1029–1120	FN-III
1139–1154	Transmembrane domain

Eight N-linked glycosylation sites.

The short form has identical extracellular and transmembrane domains. The cytoplasmic domain is, however, truncated, diverging after residue 1223 and terminating after 16 further residues: MNEDGSFIGQTGRLGL.

Database accession numbers

	PIR	SWISSPROT	EMBL/GENBANK	REFERENCE
Drosophila	A32579	P20241	M28231	

Alternative forms

Two variants differing in the length of the cytoplasmic domain are generated by alternative splicing [3].

References

[1] **Grenningloh, G. et al. (1990) Cold Spring Harbor Symp. Quant. Biol. 55, 327–340.**
[2] **Hortsch, M. and Goodman, C.S. (1991) Annu. Rev. Cell Biol. 505–557.**
[3] Hortsch, M. et al. (1990) Neuron 4, 697–709.
[4] Bieber, A.J. et al. (1989) Cell 59, 447–460.

Ng-CAM (neuronal-glial cell adhesion molecule) G4, 8D9

Family
Immunoglobulin superfamily, C2 subset. Ng-CAM is structurally related to mouse L1 and shares 40% amino acid identity.

Cellular distribution
Ng-CAM is present on post-mitotic, premigratory neurons. On post-migratory neurons Ng-CAM is largely restricted to axons. Ng-CAM is also present on Schwann cells [reviewed in ref.1].

Function
Ng-CAM mediates the adhesion of neurons to neurons and neurons to glia [2]. It is involved in neuronal migration, neurite fasciculation and outgrowth.

Regulation of expression
Ng-CAM is expressed during periods of neurite outgrowth and is dramatically downregulated in adult tissues [reviewed in ref.1].

Ligands
Ng-CAM binds homophilicly and to axonin-1 on neurons [3]. The glial cell ligand is unknown.

Structure [4]

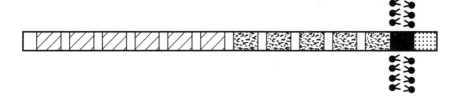

Molecular weights
Polypeptide   134 738
SDS PAGE   135 000 (predominant), 210 000, 190 000 and 210 000

Amino acid sequence (from 9–16 day embryonic brain) [4]

```
  1    MALPMVGLLL  LLLLGGPGAA  ITIPPEYGAH  DFLQPPELTE  EPPEQLVVFP

 51    SDDIVLKCVA  TGNPPVQYRW  SREISPSSPR  STGGSRWSPD  RHLVINATLA

101    ARLQGRFRCF  ATNALGTAVS  PEANVIAENT  PQWPKKKVTP  VEVEEGDPVV

151    LPCDPPESAV  PPKIYWLNSD  IVHIAQDERV  SMGQDGNLYF  SNAMVGDSHP
```

Ng-CAM

```
 201   DYICHAHFLG PRTIIQKEPL DLRVAPSNAV RSRRPRLLLP RDPQTTTIAL
                                      *
 251   RGGSVVLECI AEGLPTPWVR WRRLNGPLLP GGVGNFNKTL RLWGVTESDD
 301   GEYECVAENG RGTARGTHSV TVEAAPYWVR RPQSGVFGPG ETARLDCEVG
                                               *
 351   GKPRPQIQWS INGVPIEAAG AERRWLRGGA LVLPELRPND SAVLQCEARN
                                      *
 401   RHGPLLANAF LHVVELPLRM LTADEQRYEV VENQTVFLHC RTFGAPAPNV
                             *                             *
 451   EWLTPTLEPA LQDDRSFVFT NGSLRVSAVR GGDGGVYTCM AQNAHSNGSL
 501   TALLEVRAPT RISAPPRSAT AKKGETVTFH CGATFDPAVT PGELRWLRGG
 551   QPLPDDPRYS VAAEMTVSNV DYGDEGTIQC RASTPLDSAE AEAQLRVVGR
 601   PPSRDLQVME VDEHRVRLSW TPGDDHNSPI EKFVVEEEEE REDLQRGFGA
 651   ADVPGQPWTP PLPLSPYGRF PFRVVAVNAY GRGEHHAPSA PIETPPAAPE
                   *
 701   RNPGGVHGEG NETGNLVITW EPLPPQAWNA PWARYRVQWR PLEEPGGGGP
 751   SGGFPWAEST VDAPPVVVGG LPPFSPFQIR VQAVNGAGKG PEATPGVGHS
                             *
 801   GEDLPLVYPE NVGVELLNSS TVRVRWTLGG GPKELRGRLR GFRVLYWRLG
 851   WVGERSRRQA PPDPPQIPQS PAEDPPPFPP VALTVGGDAR GALLGGLRPW
 901   SRYQLRVLVF NGRGDGPPSE PIAFETPEGV PGPPEELRVE RLDDTALSVV
 951   ERRTFKRSIT GYVLRYQQVE PGSALPGGSV LRDPQCDLRG LNARSRYRLA
1001   LPSTPRERPA LQTVGSTKPE PPSPLWSRFG VGGRGGFHGA AVEFGAAQED
                   *                   *                   *
1051   DVEFEVQFMN KSTDEPWRTS GRANSSLRRY RLEGLRPGTA YRVQFVGRNR
                   *
1101   SGENVAFWES EVQTNGTVVP QPGGGVCTKG WFIGFVSSVV LLLLILLILC
1151   FIKRSKGGKY SVKDKEDTQV DSEARPMKDE TFGEYRSLES EAEKGSASGS
1201   GAGSGVGSPG RGPCAAGSED SLAGYGGSGD VQFNEDGSFI GQYRGPGAGP
1251   GSSGPASPCA GPPLD
```

Ng-CAM

Domain structure

1–20	Signal sequence
51–113	Ig-1
146–208	Ig-2
252–309	Ig-3
340–400	Ig-4
433–493	Ig-5
524–584	Ig-6
620–695	FN-III
720–795	FN-III
824–891	FN-III
947–1004	FN-III
1052–1091	FN-III
1130–1152	Transmembrane domain

Twelve potential N-linked glycosylation sites.

Database accession numbers

	PIR	SWISSPROT	EMBL/GENBANK	REFERENCE
Chicken			X56969	4

Alternative forms

Ng-CAM is synthesized as M_r 210 000 and M_r 190 000 variants which are proteolytically cleaved to produce fragments of M_r 135 000 and an M_r 80 000. The M_r 135 000 form consists of the six amino-terminal Ig-like domains, the first two fibronectin repeats and a portion of the third. The M_r 80 000 form consists of the remaining fibronectin repeats, together with the transmembrane and cytoplasmic domains. The M_r 135 000 fragment is non-covalently associated with the membrane bound fragment [5]. The cleavage site is between Arg 858 and Gln 859.

References

[1] **Grumet, M. (1992) J. Neurosci. Res. 31, 1–3**
[2] Grumet, M. and Edelman, G. (1988) J. Cell Biol. 106, 487–503.
[3] Kuhn, T.B. et al. (1991) J. Cell Biol. 115, 1113–1126.
[4] Burgoon, M.P. et al. (1991) J. Cell Biol. 112, 1017–1029.
[5] Thiery, J-P. et al. (1985) J. Cell Biol. 100, 442–456.

P-Cadherin — Placental cadherin

Family
Cadherin.

Cellular Distribution
Early mouse embryo: extra-embryonic ectoderm, visceral endoderm, lateral plate mesoderm, notochord. Late mouse embryo: Epidermis and other epithelial tissue, pigmented retina, mesothelium. Human: some epithelial tissues in human fetus and low levels of expression in placenta [1-3]; human epidermoid cell line A431 [4].

Function
Ca^{2+} dependent homotypic adhesion. Possibly involved in connecting embryo to uterus. In common with E- and N-cadherin, P-cadherin has been shown to associate with catenins [5].

Regulation of expression
Developmentally regulated. Highly expressed in mouse placenta throughout pregnancy in contrast to human placenta where the major cadherin appears to be E-cadherin after 6 weeks of pregnancy [4,6]. Suppression of P-cadherin expression may be lost in the process of carcinogenesis.

Ligand
P-Cadherin (homotypic adhesion) is thought to occur via the HAV sequence located in the N-terminal 113 amino acids.

Gene structure

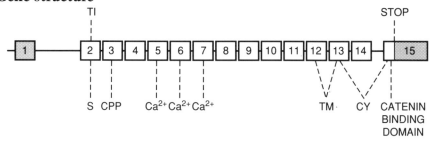

Spans 45 kb (15 exons) (mouse) [7]. The gene structure is almost identical to that found in chicken L-CAM and mouse E-cadherin except the 1st exon of P-cadherin covers the regions corresponding to exons 1 and 2 of L-CAM and E-cadherin. The first intron is 23 kb.

Gene location and size
8 (mouse); 45 kb [7], linked with E-cadherin locus.

Molecular weights
Polypeptide   79 999 (mouse)
SDS PAGE   118 000 (mouse)

P-Cadherin

Amino acid sequence (from mouse PSA5-E cells) [1]

```
                              ↓
  1  MELLSGPHAF LLLLLQVCWL RSVVSEPYRA GFIGEAGVTL EVEGTDLEPS
 51  QVLGKVALAG QGMHHADNGD IIMLTRGTVQ GGKDAMHSPP TRILRRRKRE
101  WVMPPIFVPE NGKGPFPQRL NQLKSNKDRG TKIFYSITGP GADSPPEGVF
                                                     *
151  TIEKESGWLL LHMPLDREKI VKYELYGHAV SENGASVEEP MNISIIVTDQ
201  NDNKPKFTQD TFRGSVLEGV MPGTSVMQVT ATDEDDAVNT YNGVVAYSIH
251  SQEPKEPHDL MFTIHKSTGT ISVISSGLDR EKVPEYRLTV QATDMDGEGS
301  TTTAEAVVQI LDANDNAPEF EPQKYEAWVP ENEVGHEVQR LTVTDLDVPN
351  WPAWRATYHI VGGDDGDHFT ITTHPETNQG VLTTKKGLDF EAQDQHTLYV
401  EVTNEAPFAV KLPTATATVV VHVKDVNEAP VFVPPSKVIE AQEGISIGEL
451  VCIYTAQDPD KEDQKISYTI SRDPANWLAV DPDSGQITAA GILDREDEQF
                                                     *
501  VKNNVYEVMV LATDSGNPPT TGTGTLLLTL TDINDHGPIP EPRQIIICNQ
             *          *
551  SPVPQVLNIT DKDLSPNSSP FQAQLTHDSD IYWMAEVSEK GDTVALSLKK
601  FLKQDTYDLH LSLSDHGNRE QLTMIRATVC DCHGQVFNDC PRPWKGGFIL
651  PILGAVLALL TLLLALLLLV RKKRKVKEPL LLPEDDTRDN VFYYGEEGGG
701  EEDQDYDITQ LHRGLEARPE VVLRNDVVPT FIPTPMYRPR PANPDEIGNF
751  IIENLKPANT DPTAPPYDSL MVFDYEGSGS DAASLSSLTT SASDQDQDYN
801  YLNEWGSRFK KLADMYGGGE DD
```

Domain structure

- 1–27 Signal sequence
- 28–99 Propeptide
- 100–647 Extracellular domain
- 109–207 Repeat I
- 208–320 Repeat II
- 321–432 Repeat III
- 433–538 Repeat IV
- 539–645 Repeat V
- 648–670 Transmembrane domain
- 671–822 Cytoplasmic domain

Four potential N-linked glycosylation sites in extracellular domain.
Three Ca^{2+} binding sites in extracellular domain (residues 199–203, 233–235 and 312–316).
The cytoplasmic domain contains a serine rich region for potential phosphorylation.

Database accession numbers

	PIR	SWISSPROT	EMBL/GENBANK	REFERENCE
Human	A33659		X63629	
			M81779	
			J04891	
		P22223		4
Mouse	S03163		X06340	
		P10287		1
Bovine, partial sequence		P19535		

References
[1] **Nose, A. et al. (1987) EMBO J. 6, 3655–3661.**
[2] Nose, A. and Takeichi, M. (1986) J. Cell Biol. 103, 2649–2658.
[3] Takeichi, M. (1988) Development 102, 639–655.
[4] Shimoyama, Y. et al. (1989) J. Cell Biol. 109, 1787–1794.
[5] Wheelock, M.J. and Knudsen, K.A. (1991) In vivo 5, 505–514.
[6] Kadokawa, Y. et al. (1989) Dev. Growth and Differ. 31, 23–30.
[7] Hatta, M. et al. (1991) Nucleic Acids Res. 19, 4437–4441.

PECAM-1

(platelet endothelial cell adhesion molecule) CD31, hec7, endoCAM

Family
Immunoglobulin superfamily, C2 subset.

Cellular distribution
Platelets, endothelial cells, monocytes, granulocytes and some T cells [1,6].

Function
The *in vivo* function of CD31 is uncertain. CD31 expressed in mouse L cells, induces aggregation [2]. CD31 is strongly expressed on T-cells adhering to endothelial cells suggesting a role in recruitment and transmigration [3]. On platelet activation the cytoplasmic domain of CD31 is highly phosphorylated and becomes associated with the cytoskeleton [7]. The interaction of CD31 is able to amplify integrin mediated adhesion of CD31 positive-T cells.

Regulation of expression
Constitutively expressed. Downregulated on activated granulocytes treated with fMLP [2] and on T cells following activation [4].

Ligand
PECAM-1 mediated binding is homophilic [2], although a heterophilic ligand has also been identified [5].

Structure [6]

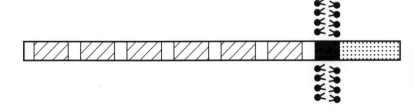

Molecular weights
Polypeptide   79 578
SDS PAGE   120 000–130 000

PECAM-1

Amino acid sequence (from endothelial cells) [6]

```
  1  MQPRWAQGAT MWLGVLLTLL LCSSLEGQEN SFTINSVDMK SLPDWTVQNG
                                  ↓
                *                            *
 51  KNLTLQCFAD VSTTSHVKPQ HQMLFYKDDV LFYNISSMKS TESYFIPEVR

101  IYDSGTYKCT VIVNNKEKTT AEYQLLVEGV PSPRVTLDKK EAIQGGIVRV
                *
151  NCSVPEEKAP IHFTIEKLEL NEKMVKLKRE KNSRDQNFVI LEFPVEEQDR

201  VLSFRCQARI ISGIHMQTSE STKSELVTVT ESFSTPKFHI SPTGMIMEGA

251  QLHIKCTIQV THLAQEFPEI IIQKDKAIVA HNRHGNKAVY SVMAMVEHSG
         *              *                              *
301  NYTCKVESSR ISKVSSIVVN ITELFSKPEL ESSFTHLDQG ERLNLSCSIP
              *
351  GAPPANFTIQ KEDTIVSQTQ DFTKIASKSD SGTYICTAGI DKVVKKSNTV

401  QIVVCEMLSQ PRISYDAQFE VIKGQTIEVR CESISGTLPI SYQLLKTSKV
           *
451  LENSTKNSND PAVFKDNPTE DVEYQCVADN CHSHAKMLSE VLRVKVIAPV

501  DEVQISILSS KVVESGEDIV LQCAVNEGSG PITYKFYREK EGKPFYQMTS
          *
551  NATQAFWTKQ KASKEQEGEY YCTAFNRANH ASSVPRSKIL TVRVILAPWK

601  KGLIAVVIIG VIIALLIIAA KCYFLRKAKA KQMPVEMSRP AVPLLNSNNE

651  KMSDPNMEAN SHYGHNDDVR NHAMKPINDN KEPLNSDVQY TEVQVSSAES

701  HKDLGKKDTE TVYSEVRKAV PDAVESRYSR TEGSLDGT
```

Domain structure

   1–27   Signal sequence
  50–113   Ig-1
 145–210   Ig-2
 249–308   Ig-3
 340–390   Ig-4
 424–480   Ig-5
 516–576   Ig-6
 602–620   Transmembrane domain

Nine potential N-linked glycosylation sites.
Potential tyrosine phosphorylation site at residue 713.

Database accession numbers

	PIR	SWISSPROT	EMBL/GENBANK
Human	A40096	P16284	M28526

Alternative forms

None known.

References
1 Stockinger, H. et al. (1990) J. Immunol. 145, 3889–3897.
2 **Albelda, S.M. et al. (1991) J. Cell Biol. 114, 1059–1068.**
3 Bogen, S.A. (1992) Am. J. Pathol. 141, 843–845.
4 Zehnder, J.L. et al. (1992) J. Biol. Chem. 267, 5243–5249.
5 Muller, W.A. et al. (1992) J. Exp. Med 175, 1401–1404.
6 Newman, P.J. et al. (1990) Science 247, 1219–1222.
7 Newman, P.J. et al. (1992) J. Cell. Biol. 119, 239–246.
8 Yoshiga, T. et al (1992) J. Exp. Med. 176, 245–253.

P-Selectin

GMP-140, CD62, LECAM-3, PADGEM

Family
Selectin, LECAM.

Cellular distribution
Platelets, endothelial cells, megakaryocytes.

Function
P-Selectin mediates the adhesion of neutrophils and monocytes to activated platelets and endothelial cells [reviewed in refs. 1,2]. Soluble P-selectin binds to a subpopulation of T-cells, indicating a potential role in T-cell recruitment at sites of inflammation [12].

Regulation of expression
P-Selectin is translocated to the cell surface within minutes, from α granules of platelets or Weibel-Palade bodies of endothelial cells, following stimulation with thrombin, histamine, PMA or peroxides [1,3].

Ligands
P-Selectin interacts with sialylated, fucosylated lactosaminoglycans, including the sialyl Lewis^x moiety on neutrophils [4,5]. A recombinant P-selectin-immunoglobulin fusion protein binds to sulphatides [6]. A potential glycoprotein ligand has been identified [7]. Adhesion is calcium-dependent.

Gene structure [8]

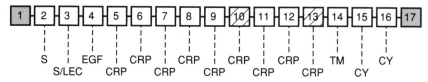

Gene location and size [9]
1, bands 21–24; >50 kb.

Structure [10]

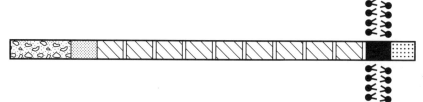

Molecular weights
Polypeptide   86 243
SDS PAGE   140 000

P-Selectin

Amino acid sequence [10]

```
  1  MANCQIAILY QRFQRVVFGI SQLLCFSALI SELTNQKEVA A↓WTYHYSTKA
                *                                              *
 51  YSWNISRKYC QNRYTDLVAI QNKNEIDYLN KVLPYYSSYY WIGIRKNNKT
101  WTWVGTKKAL TNEAENWADN EPNNKRNNED CVEIYIKSPS APGKWNDEHC
                                   *
151  LKKKHALCYT ASCQDMSCSK QGECLETIGN YTCSCYPGFY GPECEYVREC
                      *          *
201  GELELPQHVL MNCSHPLGNF SFNSQCSFHC TDGYQVNGPS KLECLASGIW
251  TNKPPQCLAA QCPPLKIPER GNMICLHSAK AFQHQSSCSF SCEEGFALVG
301  PEVVQCTASG VWTAPAPVCK AVQCQHLEAP SEGTMDCVHP LTAFAYGSSC
351  KFECQPGYRV RGLDMLRCID SGHWSAPLPT CEAISCEPLE SPVHGSMDCS
                     *
401  PSLRAFQYDT NCSFRCAEGF MLRGADIVRC DNLGQWTAPA PVCQALQCQD
                   *
451  LPVPNEARVN CSHPFGAFRY QSVCSFTCNE GLLLVGASVL QCLATGNWNS
                     *
501  VPPECQAIPC TPLLSPQNGT MTCVQPLGSS SYKSICQFIC DEGYSLSGPE
551  RLDCTRSGRW TDSPPMCEAI KCPELFAPEQ GSLDCSDTRG EFNVGSTCHF
601  SCNNGFKLEG PNNVECTTSG RWSATPPTCK GIASLPTPGL QCPALTTPGO
                     *
651  GTMYCRHHPG TFGFNTTCYF GCNAGFTLIG DSTLSCRPSG QWTAVTPACR
                *           *                         *
701  AVKCSELHVN KPIAMNCSNL WGNFSYGSIC SGHCLEGQLL NGSAQTACQE
751  NGHWSTTVPT CQAGPLTIQE ALTYFGGAVA STIGLIMGGT LLALLRKRFR
801  QKDDGKCPLN PHSHLGTYGV FTNAAFDPSP
```

Domain structure
    1–41   Signal sequence
   42–160   Lectin domain
  161–199   EGF domain
  200–257   CRP domain
  262–319   CRP domain
  324–381   CRP domain
  386–443   CRP domain
  448–505   CRP domain
  510–567   CRP domain
  572–629   CRP domain
  642–699   CRP domain
  704–761   CRP domain
  772–795   Transmembrane domain

Twelve potential N-linked glycosylation sites.

Deletion of residues 569–630 (exon 11) deletes CRP-domain 7. Deletion of residues 763–802 (exon 14) deletes the transmembrane domain [10].

Database accession numbers

	PIR	SWISSPROT	EMBL/GENBANK	REFERENCE
Human	A30359	P16109	M25322	
Mouse			M87861	13

Alternative forms

cDNA clones have been isolated which predict the existence of two variants produced by alternative splicing. One of these is missing the exon encoding CRP-domain 7. The other is missing the exon encoding the transmembrane domain and is predicted to be soluble [10]. A soluble form of P-selectin has been detected in serum [11].

References
[1] **McEver, R.P. (1991) J. Cell. Biochem. 45, 156–161.**
[2] Lawrence, M.B. and Springer, T.A. (1991) Cell 65, 859–873.
[3] Lorant, D.E. et al. (1991) J. Cell Biol. 115, 223–234.
[4] Polley, M.J. et al. (1991) Proc. Natl Acad. Sci. USA 88, 6226–6228.
[5] Zhou, Q. et al. (1991) J. Cell Biol. 115, 557–564.
[6] Aruffo, A. et al. (1991) Cell 67, 35–44.
[7] Moore, K.L. et al. (1992) J. Cell Biol. 118, 445–456.
[8] Johnston, G.I. et al. (1990) J. Biol.Chem. 265, 21381–21385.
[9] Watson, M.L. et al. (1990) J. Exp. Med. 172, 263–272.
[10] Johnston, G.I. et al. (1989) Cell 556, 1033–1044.
[11] Dunlop, L.C. et al. (1992) J. Exp. Med. 175, 1147–1150.
[12] Moore, K.L. and Thompson, L.F. (1992) Biochem. Biophys. Res. Comm. 186, 173–181.
[13] Weller, A. et al. (1992) J. Biol. Chem. 267, 15176–15183.

Platelet glycoprotein GPIIb-IIIa

Integrin αIIb/ β3, CD41/CD61

Family
β3 integrin (dimer of αIIb and β3 subunits).

Cellular distribution
Platelets, megakaryoblasts.

Function
Platelet aggregation. GPIIb–IIIa is also a receptor for several soluble adhesive proteins, the binding of which is mediated in part by RGD recognition sequences. Tyrosine phosphorylation of several intracellular proteins can be induced by GPIIb–IIIa [1,2].

Regulation of expression
Constitutively expressed. On unstimulated platelets, GPIIb–IIIa is randomly dispersed on the surface and is only capable of recognizing immobilized fibrinogen but not other soluble RGD-containing proteins such as vWF and vitronectin [3].

On activated platelets (triggered by thrombin, collagen or ADP), the receptor function of GPIIb–IIIa becomes activated causing the acquisition of high affinity Fg and vWF binding, allowing for aggregate formation [4]. The cytoplasmic domain of GPIIb may be important in the control of ligand binding affinity [5]. Activation is accompanied by a conformational change in GPIIb–IIIa. A further conformational change occurs on ligand binding [6].

In an autosomal recessive disorder, Glanzmann thrombasthaemia, GPIIb–IIIa is either missing or the receptor function of GPIIb–IIIa is defective (platelets do not bind adhesive proteins, therefore fail to aggregate resulting in bleeding diathesis) [7].

Ligands
Fibrinogen, fibronectin, vWF, vitronectin and thrombospondin [8].

GPIIb–IIIa recognizes an RGD sequence in all ligands although a KQAGDV sequence has been identified in the fibrinogen γ chain.

Residues 237–248 of GPIIIa (β3) are critical in adhesive protein binding [9].

Gene structure
GPIIb (αIIβ) spans 17.2 kb and contains 30 exons whose demarcations do not correlate with suggested functional domains [10]. The 5' flanking sequence does not contain obvious TATA or CAAT boxes.

Platelet glycoprotein GPIIb-IIIa

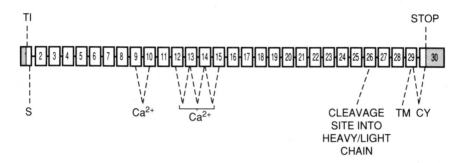

GPIIIa (β3) is encoded by 14 exons spanning 60 kb [11,12].

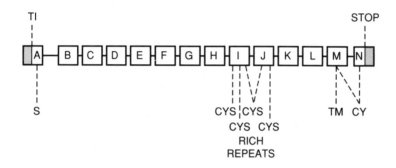

Gene location
Both subunits map to chromosome 17q21–23 [17].

Structure

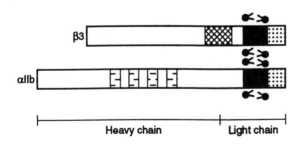

Molecular weights
Polypeptide   110 019 (αIIb); 87 213 (β3)
SDS PAGE   123 000/23 000 (αIIb); 105 000 (β3)

Platelet glycoprotein GPIIb-IIIa

Amino acid sequence (αIIb) (from HEL cells) [14]

```
  1  MARALCPLQA LWLLEWVLLL LGPCAAPPAW ALNLDPVQLT FYAGPNGSQF
 51  GFSLDFHKDS HGRVAIVVGA PRTLGPSQEE TGGVFLCPWR AEGGQCPSLL
101  FDLRDETRNV GSQTLQTFKA RQGLGASVVS WSDVIVACAP WQHWNVLEKT
151  EEAEKTPVGS CFLAQPESGR RAEYSPCRGN TLSRIYVEND FSWDKRYCEA
201  GFSSVVTQAG ELVLGAPGGY YFLGLLAQAP VADIFSSYRP GILLWHVSSQ
251  SLSFDSSNPE YFDGYWGYSV AVGEFDGDLN TTEYVVGAPT WSWTLGAVEI
301  LDSYYQRLHR LRAEQMASYF GHSVAVTDVN GDGRHDLLVG APLYMESRAD
351  RKLAEVGRVY LFLQPRGPHA LGAPSLLLTG TQLYGRFGSA IAPLGDLDRD
401  GYNDIAVAAP YGGPSGRGQV LVFLGQSEGL RSRPSQVLDS PFPTGSAFGF
451  SLRGAVDIDD NGYPDLIVGA YGANQVAVYR AQPVVKASVQ LLVQDSLNPA
501  VKSCVLPQTK TPVSCFNIQM CVGATGHNIP QKLSLNAELQ LDRQKPRQGR
551  RVLLLGSQQA GTTLNLDLGG KHSPICHTTM AFLRDEADFR DKLSPIVLSL
601  NVSLPPTEAG MAPAVVLHGD THVQEQTRIV LDCGEDDVCV PQLQLTASVT
651  GSPLLVGADN VLELQMDAAN EGEGAYEAEL AVHLPQGAHY MRALSNVEGF
701  ERLICNQKKE NETRVVLCEL GNPMKKNAQI GIAMLVSVGN LEEAGESVSF
751  QLQIRSKNSQ NPNSKIVLLD VPVRAEAQVE LRGNSFPASL VVAAEEGERE
801  QNSLDSWGPK VEHTYELHNN GPGTVNGLHL SIHLPGQSQP SDLLYILDIQ
851  PQGGLQCFPQ PPVNPLKVDW GLPIPSPSPI HPAHHKRDRR QIFLPEPEQP
901  SRLQDPVLVS CDSAPCTVVQ CDLQEMARGQ RAMVTVLAFL WLPSLYQRPL
951  DQFVLQSHAW FNVSSLPYAV PPLSLPRGEA QVWTQLLRAL EERAIPIWWV
1001 LVGVLGGLLL LTILVLAMWK VGFFKRNRPP LEEDDEEGE
```

Domain structure

The α subunit is synthesized as a single chain precursor which undergoes cleavage to produce a disulphide-linked heavy and light chain. There are four divalent metal ion binding sites in this subunit.

Platelet glycoprotein GPIIb-IIIa

Ca2+ is necessary for heterodimer formation between αIIb and β3 [14].

1–31	Signal sequence
32–902	Heavy chain
903–1039	Light chain
994–1019	Transmembrane domain
1020–1039	Cytoplasmic domain

Five potential N-linked glycosylation sites.

Amino acid sequence (β3) (from HUVECs) [15]

```
  1   MRARPRPRPL WVTVLALGAL AGVGVGGPNI CTTRGVSSCQ QCLAVSPMCA
 51   WCSDEALPLG SPRCDLKENL LKDNCAPESI EFPVSEARVL EDRPLSDKGS
101   GDSSQVTQVS PQRIALRLRP DDSKNFSIQV RQVEDYPVDI YYLMDLSYSM
151   KDDLWSIQNL GTKLATQMRK LTSNLRIGFG AFVDKPVSPY MYISPPEALE
201   NPCYDMKTTC LPMFGYKHVL TLTDQVTRFN EEVKKQSVSR NRDAPEGGFD
251   AIMQATVCDE KIGWRNDASH LLVFTTDAKT HIALDGRLAG IVQPNDGQCH
301   VGSDNHYSAS TTMDYPSLGL MTEKLSQKNI NLIFAVTENV VNLYQNYSEL
351   IPGTTVGVLS MDSSNVLQLI VDAYGKIRSK VELEVRDLPE ELSLSFNATC
401   LNNEVIPGLK SCMGLKIGDT VSFSIEAKVR GCPQEKEKSF TIKPVGFKDS
451   LIVQVTFDCD CACQAQAEPN SHRCNNGNGT FECGVCRCGP GWLGSQCECS
501   EEDYRPSQQD ECSPREGQPV CSQRGECLCG QCVCHSSDFG KITGKYCECD
551   DFSCVRYKGE MCSGHGQCSC GDCLCDSDWT GYYCNCTTRT DTCMSSNGLL
601   CSGRGKCECG SCVCIQPGSY GDTCEKCPTC PDACTFKKEC VECKKFDREP
651   YMTENTCNRY CRDEIESVKE LKDTGKDAVN CTYKNEDDCV VRFQYYEDSS
701   GKSILYVVEE PECPKGPDIL VVLLSVMGAI LLIGLAALLI WKLLITIHDR
751   KEFAKFEEER ARAKWDTANN PLYKEATSTF TNITYRGT
```

Domain structure

The β3 chain contains four tandem cysteine-rich repeats.

1–26	Signal sequence
27–788	Extracellular domain

719–741	Transmembrane domain	
742–788	Cytoplasmic domain	
463–511	Cysteine-rich repeat	
512–553	Cysteine-rich repeat	
554–592	Cysteine-rich repeat	
593–629	Cysteine-rich repeat	

Tyr $_{773}$ is a potential phosphorylation site.

Six potential N-linked glycosylation sites.

Database accession numbers

	PIR	SWISSPROT	EMBL/GENBANK
Human αIIβ	A28937	P08514	J02764
Human β3	A02109	P05106	J02703

Alternative forms

An alternatively spliced β3 cDNA has been isolated from a placental cDNA library. This variant contains a cytoplasmic domain which is eight amino acids shorter than the usual β3 subunit, and has a different 13 amino acid C-terminal peptide [16]. The novel cytoplasmic domain also lacks a tyrosine residue. This mRNA arises due to defective splicing occurring between exons M and N.

αIIb can be alternatively spliced in platelets and megakaryoblasts yielding a variant with a 102 bp deletion (corresponding to a 34 amino acid deletion) in the extracellular domain [13].

References

[1] Ferrell, J.E. and Martin, G.S. (1989) Proc. Natl Acad. Sci. USA 86, 2234–2238.
[2] Golden, A. et al. (1990) J. Cell Biol. 111, 3117–3127.
[3] Kieffer, N. et al. (1991) J. Cell Biol. 113, 451–461.
[4] Du, X. et al. (1991) Cell 65, 409–416.
[5] O'Toole, T. E. et al. (1991) Science 254, 845–847.
[6] Frelinger, A.L. et al. (1991) J. Biol. Chem. 266, 17106–17111.
[7] Kieffer, N. and Phillips, D.R. (1990) Annu. Rev. Cell Biol. 6, 329–357.
[8] **Phillips, D.R. et al. (1991) Cell 65, 359–362.**
[9] Charo, I.F. et al. (1991) J. Biol. Chem. 266, 1415–1421.
[10] Heidenreich, R. et al. (1990) J. Biol. Chem. Bioch. 29, 1232–1244.
[11] Zimrin, A.B. et al. (1990) J. Biol. Chem. 265, 8590–8595.
[12] Lanza, F. et al. (1990) J. Biol. Chem. 265, 18098–18103.
[13] Bray, P.F. et al. (1990) J. Biol. Chem. 265, 9587–9590.
[14] Poncz, M. et al. (1987) J. Biol. Chem. 262, 8476–8482.
[15] Fitzgerald, L.A. et al. (1987) J. Biol. Chem. 262, 3936–3939.
[16] van Kuppevelt, T. et al. (1989) Proc. Natl Acad. Sci. USA 5415–5418.
[17] Sosnoski, D.M. et al. (1988) J. Clin. Invest. 81, 1993–1998.

Platelet Glycoprotein GPIb–IX complex

GPIb (CD42b) and GPIX (CD42a)

Family
Leucine-rich repeat family.

Cellular distribution
Platelets, megakaryoblasts. GPIb has also been reported on vascular and tonsillar endothelial cells [1].

Function
For recent reviews, see references 2–4.

GPIb–IX mediates platelet adhesion to damaged blood vessels through an interaction with subendothelial von Willebrand factor (vWF). Also acts as a thrombin receptor, thereby modulating the platelet response to thrombin. The GPIb–IX complex serves as a site of attachment to the platelet membrane skeleton through actin binding protein. This interaction may be important in maintaining platelet shape and for regulating the function of other platelet membrane glycoproteins.

Recently the GPIb–IX complex has been shown to associate with another member of the leucine rich repeat family, platelet glycoprotein GPV [5]. The role of GPV is as yet unknown.

Regulation of expression
Constitutively expressed. All three subunits are required for plasma membrane expression. Thrombin can down regulate surface expression of the GPIb–IX complex [6]. In endothelial cells expression of GPIb can be induced by TNF or IFNγ [1]. An inherited deficiency or defect in GPIb expression causes several bleeding disorders including Bernard-Soulier syndrome [7]. (GPV is also absent in Bernard-Soulier syndrome).

Ligand
vWF and thrombin (1 high and 1 low affinity site located in the GPIbα chain).

Gene structure

GPIbα spans 6kb and contains two exons. The single intron of 233 bp occurs 6 bp upstream of the initiation site. The gene also contains five *Alu* repeats in the 5' flanking region and two *Alu* repeats in the 3' flanking region. An inverted AP-1 binding site and two AP2 binding sites have also been detected [8,9].

Gene location
17p12-ter (GPIbα) [8]; 22 (GPIbβ) [10]; 3 (GPIX) [12],

Structure
Platelet glycoprotein GPIb–IX consists of a GPIbα chain linked to a GPIbβ chain by a disulphide bridge [11]. This forms a 1:1 complex (non-covalent) with GPIX [12].

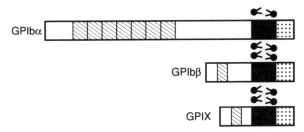

Molecular weights
Polypeptide   67 192 kD (GPIbα); 19 216 kD (GPIbβ); 17 258 kD (GPIX)
SDS PAGE   145 kD (GPIbα); 25 kD (GPIbβ); 20kD (GPIX)

Amino acid sequence (GPIbα) (from PMA-induced human erythroleukaemia cell line (HEL)) [17]

```
  1   MPLLLLLLLL PSPLHPHPIC EVSKVASHLE VNCDKRNLTA LPPDLPKDTT
 51   ILHLSENLLY TFSLATLMPY TRLTQLNLDR CELTKLQVDG TLPVLGTLDL
101   SHNQLQSLPL LGQTLPALTV LDVSFNRLTS LPLGALRGLG ELQELYLKGN
151   ELKTLPPGLL TPTPKLEKLS LANNNLTELP AGLLNGLENL DTLLLQENSL
201   YTIPKGFFGS HLLPFAFLHG NPWLCNCEIL YFRRWLQDNA ENVYVWKQGV
251   DVKAMTSNVA SVQCDNSDKF PVYKYPGKGC PTLGDEGDTD LYDYYPEEDT
301   EGDKVRATRT VVKFPTKAHT TPWGLFYSWS TASLDSQMPS SLHPTQESTK
351   EQTTFPPRWT PNFTLHMESI TFSKTPKSTT EPTPSPTTSE PVPEPAPNMT
401   TLEPTPSPTT PEPTSEPAPS PTTPEPTPIP TIATSPTILV SATSLITPKS
451   TFLTTTKPVS LLESTKKTIP ELDQPPKLRG VLQGHLESSR NDPFLHPDFC
501   CLLPLGFYVL GLFWLLFASV VLILLLSWVG HVKPQALDSG QGAALTTATQ
551   TTHLELQRGR QVTVPRAWLL FLRGSLPTFR SSLFLWVRPN GRVGPLVAGR
601   RPSALSQGRG QDLLSTVSIR YSGHSL
```

Platelet Glycoprotein GPIb–IX complex

Domain structure

1–16	Signal sequence
17–501	Extracellular domain
52–62	Leucine rich repeat
63–86	Leucine rich repeat
87–108	Leucine rich repeat
109–131	Leucine rich repeat
132–155	Leucine rich repeat
156–179	Leucine rich repeat
180–202	Leucine rich repeat
378–386	Threonine rich repeat
387–399	Threonine rich repeat
400–408	Threonine rich repeat
409–421	Threonine rich repeat
422–430	Threonine rich repeat
502–530	Transmembrane domain
531–626	Cytoplasmic domain

Four potential N-linked glycosylation sites in extracellular domain.

One or both cysteine residues located immediately before the transmembrane domain may be involved in disulphide linkage to GPIb(β).

vWF binding site is located in the N-terminal 45 kD.

The cytoplasmic domain interacts with actin binding protein [14].

The extracellular domain contains 7 homologous leucine rich repeats and 5 threonine rich repeats. The threonine rich repeats are subject to 'O' glycosylation [15,16].

Amino acid sequence (GPIbβ) (from HEL cells [18])

```
                                    ↓
  1   MGSGPRGALS LLLLLLAPPS RPAAGCPAPC SCAGTLVDCG RRGLTWASLP
                      *
 51   TAFPVDTTEL VLTGNNLTAL PPGLLDALPA LRTAHLGANP WRCDCRLVPL

101   RAWLAGRPER APYRDLRCVA PPALRGRLLP YLAEDELRAA CAPGPLCWGA

151   LAAQLALLGL GLLHALLLVL LLCRLRRLRA RARARAAARL SLTDPLVAER

201   AGTDES
```

Domain structure

1–25	Signal sequence
26–147	Extracellular domain
60–83	Leucine rich repeat
148–172	Transmembrane domain
173–206	Cytoplasmic domain

The cytoplasmic domain contains serine and threonine residues which can be phosphorylated by cAMP dependent protein kinase and which may be involved in the regulation of actin polymerization [19]. This domain also contains an active thiol group which is partly palmitoylated [20].

A leucine rich sequence homologous to that found in the α chain occurs between residues 60 and 83.

One potential N-linked glycosylation site (of the biantennary lactosamine type) in the extracellular domain [21].

Amino acid sequence (GPIX) (from HEL cells [12])

```
              ↓
  1   MPAWGALFLL WATAEATKDC PSPCTCRALE TMGLWVDCRG HGLTALPALP
                  *
 51   ARTRHLLLAN NSLQSVPPGA FDHLPQLQTL DVTQNPWHCD CSLTYLRLWL

101   EDRTPEALLQ VRCASPSLAA HGPLRLTGYQ LGSCGWQLQA SWVRPGVLWD

151   VALVAVAALG LALLAGLLCA TTEALD
```

Domain structure
- 1–16   Signal sequence
- 17–145 Extracellular domain
- 56–79   Leucine rich homology to α and β chains
- 146–170 Transmembrane domain
- 171–176 Cytoplasmic tail

One potential N-linked glycosylation site (of the biantennary lactosamine type) in the extracellular domain [21].

Cytoplasmic domain is palmitoylated [20].

Database accession numbers

	PIR	SWISSPROT	EMBL/GENBANK
Human GPIbα	A27075	P07359	J02940
Human GPIbβ	A31929	P13224	J03259
Human GPIX	A33731	P14770	X52997

Alternative forms

Following platelet lysis *in vitro* or platelet destruction *in vivo*, the extracellular portion of GPIbα is cleaved from the platelet surface yielding a water soluble fragment known as glycocalicin [22].

Polymorphisms of GPIb are also known. Variation in 'O' glycosylation due to duplication, triplication or quadruplication of a 13 amino acid sequence cassette yields types A, B, C and D which differ in molecular weight (D is the normal form) [23]. A dimorphism arising from a point mutation has also been detected [24].

References

[1] Rajagopalan, V. et al. (1992) Blood 80, 153–161.
[2] Clemetson, K.J. et al. (1990) Prog. Clin. Biol. Res. 356, 77–88.
[3] Ruggieri, Z.M. (1991) Prog. Hemost. Thromb. 10, 35–68.
[4] Roth, G.J. (1991) Blood 77, 5–19.
[5] Modderman, P.W. et al. (1992) J. Biol. Chem. 267, 364–369.
[6] Michelson, A.D. et al. (1991) Blood 77, 770–779.
[7] Clemetson, K.J. and Luscher, E.F. (1988) Biochim. Biophys. Acta 947, 53–73.
[8] Wenger, R.H. et al. (1989) Gene 85, 517–524.
[9] Wenger, R.H. et al. (1988) Biochem. Biophys. Res. Commun. 156, 389–395.
[10] Bennett, J.S. (1990) Semin. Hematol. 27, 186–204.
[11] Phillips, D.R. and Poh-Agin, P. (1977) J. Biol. Chem. 252, 2121–2126.
[12] Hickey, M. J. et al. (1990) FEBS letts. 274, 189–192.
[13] Berndt, M.C. et al. (1985) Eur. J. Biochem. 151, 637–649.
[14] Fox, J.E.B. et al. (1985) J. Biol. Chem. 260, 11970–11977.
[15] Korrel, S.A.M. et al. (1984) Eur. J. Biochem. 140, 571–576.
[16] Tsuji, T. et al. (1983) J. Biol. Chem. 258, 6335–6339.
[17] Lopez, J.A. et al. (1987) Proc. Natl Acad. Sci. USA 84, 5615–5619.
[18] Lopez, J.A. et al. (1988) Proc. Natl Acad. Sci. USA 85, 2135–2139.
[19] Fox, J.E.B. et al. (1989) J. Biol. Chem. 264, 9520–9526.
[20] Muszbek, L. and Laposata M. (1989) J. Biol. Chem. 264, 9716–9719.
[21] Wicki, A.N and Clemetson K.J. (1987) Eur. J. Biochem. 163, 43–50.
[22] Clemetson, K.J. et al. (1981) Proc. Natl Acad. Sci. USA 78, 2712–2716.
[23] Lopez, J.A. et al. (1992) J. Biol. Chem. 267, 10055–10061.
[24] Murata, M. et al. (1992) Blood 79, 3086–3090.

R-Cadherin — Retinal Cadherin

Family
Cadherin.

Cellular distribution
Chicken retina [1].

Function
Ca^{2+} dependent homotypic adhesion of retinal cells, thus playing an important role in retinal development [1].

Regulation of expression
Expression is restricted to neural retinal cells and appears only after some degree of neuronal differentiation has taken place (around embryonic day 8 – later than N-cadherin). It is expressed in the outer layer of the optic cup that gives rise to the optic stalk and extends to pigmented retina. Maximal expression occurs on day 14 and persists until hatching [1]. In general, expression is complementary to that of N-cadherin [2].

Ligand
R-Cadherin (homotypic adhesion); however, some binding to N-cadherin has been observed [1].

Gene structure
Unknown.

Chromosomal location
Unknown.

Molecular weights
Polypeptide	82 200
SDS PAGE	124 000

R-Cadherin

Amino acid sequence (from chicken embryonic retina) [1]

```
  1  MRTGSRLLLV LLVWGSAAAL NGDLTVRPTC KPGFSEEDYT AFVSQNIMEG
 51  QKLLKVKFNN CAGNKGVRYE TNSLDFKVRA DGTMYAVHQV QMASKQLILM
101  VTAWDPQTLG RWEAIVRFLV GEKLQHNGHK PKGRKSGPVD LAQQQSDTLL
151  PWRQHQSAKG LRRQKRDWVI PPINVPENSR GPFPQQLVRI RSDKDKEIHI
201  RYSITGVGAD QPPMEVFSID PVSGRMYVTR PMDREERASY HLRAHAVDMN
251  GNKVENPIDL YIYVIDMNDN RPEFINQVYN GSVDEGSKPG TYVMTVTAND
301  ADSTTANGM  VRYRIVTQTP QSPSQNMFTI NSETGDIVTV AAGLDREKVQ
351  QYMVIVQATD MEGNLNYGLS NTATAIITVT DVNDNPPEFT TSTYSGEVPE
401  NRVEVVVANL TVMDRDQPHS PNWNAIYRII SGDPSGHFTI RTDPVTNEGM
451  VTVVKAVDYE MNRAFMLTVM VSNQAPLASG IQMSFQSTAG VTISVTDVNE
501  APYFPTNHKL IRLEEGVPTG TVLTTFSAVD PDRFMQQAVR YSKLSDPANW
551  LNINATNGQI TTAAVLDRES DYIKNNVYEA TFLAADNGIP PASGTGTLQI
601  YLIDINDNAP ELLPKEAQIC EKPNLNVINI TAADADIDPN VGPFVFELPS
651  VPSAVRKNWT ITRLNGDYAQ LSLRIMYLEA GVYDVPIIVT DSGNPPLYNT
701  SIIKVKVCPC DENGDCTTIG AVAAAGLGTG AIIAILICII ILLTMVLLFV
751  VWMKRREKER HTKQLLIDPE DDVRDNILKY DEEGGGEEDQ DYDLSQLQQP
801  ETMDHVLNKA PGVRRVDERP IGAEPQYPIR PVIPHPGDIG DFINEGLRAA
851  DNDPTAPPYD SLLVFDYEGS GSTAGSVSSL NSSSSGDQDY DYLNDWGPRF
901  KKLADMYGGG EED
```

A human homologue of chicken R-cadherin has recently been identified in human adult and fetal brain [3].

R-Cadherin

Domain structure

Highly homologous to N-cadherin.

1–26	Signal sequence
27–166	Propeptide
167–722	Extracellular domain
167–274	Repeat I
275–389	Repeat II
390–504	Repeat III
505–610	Repeat IV
611–721	Repeat V
723–753	Transmembrane domain
756–913	Cytoplasmic domain

Five potential N-linked glycosylation sites in the extracellular domain.

Three Ca^{2+} binding sites in the extracellular domain (residues 266–270, 299–302 and 361–365).

RQKR sequence at position 162 is the recognition site for protease cleavage to give the mature form [4].

Database accession numbers

	PIR	SWISSPROT	EMBL/GENBANK	REFERENCE
Chicken			D00849	1

Alternative forms

Unknown.

References

[1] **Inuzuka, H. et al. (1991) Neuron 7, 69–79.**
[2] Ranscht, B. (1991) Seminars Neurosci. 3, 285–296.
[3] Suzuki, S. et al. (1991) Cell Regulation 2, 261–270.
[4] Ozawa, M. and Kemler, R. (1990) J. Cell Biol. 111, 1645–1650.

T-Cadherin — Truncated cadherin

Family
Cadherin.

Cellular distribution

Neural tissue (T-cadherin is distributed on nerve fibres of lateral and ventral white matter, but not on dorsal longitudinal nerve fascicles). Also found in skeletal muscle, heart, kidney, retina and low levels are detectable in lung [1].

Function

Remains to be established. May be involved in Ca^{2+}-dependent homotypic and possibly heterotypic adhesion. Co-localizes with N-cadherin in some cell populations and in individual cells. May play an instructive role in axon growth and guidance in developing embryos, by influencing the pattern of neural crest cell migration and in maintaining somite polarity [1,2].

T-Cadherin differs from known cadherins in that it lacks cytoplasmic sequences, which are highly conserved in the other members of the family.

Regulation of expression

Developmentally regulated [2]. During chick embryonic development it has a spatially and temporally restricted distribution [1]. It is transiently expressed in the ventral region of the spinal cord (including the floor plate) during extension of commissural axons [3].

Ligand
Homotypic adhesion. May also bind N-cadherin. Differs from other cadherins in that binding does not appear to involve an N-terminal HAV peptide [1].

Gene structure

Unknown.

Gene location
Unknown.

Molecular weights
Polypeptide   78 388
SDS PAGE   95 000 (110 000 precursor)

T-Cadherin

Amino acid sequence (from chick brain E13 embryo) [1].

```
  1    MQHKTQLTLS FLLSQVLLLA CAEDLECTPG FQQKVFYIEQ PFEFTEDQPI
 51    LNLVFDDCKG NNKLNFEVSN PDFKVEHDGS LVALKNVSEA GRALFVHARS
101    EHAEDMAEIL IVGADEKHDA LKEIFKIEGN LGIPRQKRAI LATPILIPEN
151    QRPPFPRSVG KVIRSEGTEG AKFRLSGKGV DQDPKGIFRI NEISGDVSVT
201    RPLDREAIAN YELEVEVTDL SGKIIDGPVR LDISVIDQND NRPMFKEGPY
251    VGHVMEGSPT GTTVMRMTAF DADDPSTDNA LLRYNILKQT PTKPSPNMFY
301    IDPEKGDIVT VVSPVLLDRE TMETPKYELV IEAKDMGGHD VGLTGTATAT
351    ILIDDKNDHP PEFTKKEFQA TVKEGVTGVI VNLTVGDRDD PATGAWRAVY
401    TIINGNPGQS FEIHTNPQTN EGMLSVVKPL DYEISAFHTL LIKVENEDPL
451    IPDIAYGPSS TATVQITVED VNEGPVFHPN PMTVTKQENI PIGSIVLTVN
501    ATDPDTLQHQ TIRYSVYKDP ASWLEINPTN GTVATTAVLD RESPHVQDNK
551    YTALFLAIDS GNPPATGTGT LHITLEDVND NVPSLYPTLA KVCDDAKDLR
601    VVVLGASDKD LHPNTDPFKF ELSKQSGPEK LWRINKLNNT HAQVVLLQNL
651    KKANYNIPIS VTDSGKPPLT NNTELKLQVC SCKKSRMDCS ASDALHISMT
701    LILLSLFSLF CL
```

Domain structure

1–22	Signal peptide
23–139	Precursor
140–246	EC1
247–363	EC2
364–477	EC3
478–583	EC4
584–693	EC5
135–138	Protease cleavage recognition sequence [4].

Seven potential N-linked glycosylation sites.
Three Ca^{2+} binding sites (residues 237–241, 271–273 and 387–389).
Anchored to the neuronal plasma membrane by a GPI-linkage.

T-Cadherin

Database accession numbers
	PIR	SWISSPROT	EMBL/GENBANK
Chicken			M81779
			J04891

Alternative forms
None known.

References
1 **Ranscht, B. and Dours-Zimmermann, M.T. (1991) Neuron 7, 391–402.**
2 Ranscht, B. and Bronner-Fraser, M. (1991) Development 111, 15–22.
3 Ranscht, B. (1991) Seminars Neurosci. 3, 285–296.
4 Ozawa, M. and Kemler, R. (1990) J. Cell Biol. 111, 1645–1650.

TAG-1

Axonin-1 (chick)

Family

Immunoglobulin superfamily, C2 subset. TAG-1 is related, but not homologous, to mouse F3 and chick contactin.

Cellular distribution

Restricted to neuronal tissues in the developing embryo [2,5]. Present in adult brain and spinal cord [1].

Function

Neurons cultured on TAG-1, or axonin-1, show extensive neurite outgrowth. The distribution and similarities of TAG-1 to known cell adhesion molecules suggest a role in neurite fasciculation and outgrowth [2].

Regulation of expression

The expression of TAG-1 is developmentally regulated [3].

Ligands

Microspheres coated with TAG-1 bind to microspheres coated with G-4 (Ng-CAM) and to neurites in culture. These interactions are blocked by antibodies to G-4 [6]. There is also evidence for homophilic interactions (M. Hynes, unpublished observation).

Chromosomal location and size

Mouse chromosome 1.

Structure [2,4]

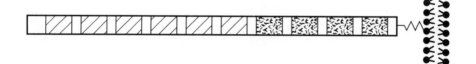

Molecular weights

Polypeptide   107 189
Reduced SDS PAGE 135 000

TAG-1

Amino acid sequence [2]

```
   1  MGTHARKKAS  LLLLVLATVA  LVSSPGWSFA  QGTPATFGPI  FEEQPIGLLF
                                      *
  51  PEESAEDQVT  LACRARASPP  ATYRWKMNGT  DMNLEPGSRH  QLMGGNLVIM
 101  SPTKTQDAGV  YQCLASNPVG  TVVSKEAVLR  FGFLQEFSKE  ERDPVKTHEG
                                                              *
 151  WGVMLPCNPP  AHYPGLSYRW  LLNEFPNFIP  TDGRHFVSQT  TGNLYIARTN
            *
 201  ASDLGNYSCL  ATSHMDFSTK  SVFSKFAQLN  LAAEDPRLFA  PSIKARFPPE
 251  TYALVGQQVT  LECFAFGNPV  PRIKWRKVDG  SLSPQWATAE  PTLQIPSVSF
 301  EDEGTYECEA  ENSKGRDTVQ  GRIIVQAQPE  WLKVISDTEA  DIGSNLRWGC
 351  AAAGKPRPMV  RWLRNGEPLA  SQNRVEVLAG  DLRFSKLSLE  DSGMYQCVAE
 401  NKHGTIYASA  ELAVQALAPD  FRQNPVRRLI  PAARGGEISI  LCQPRAAPKA
                    *                *                         *
 451  TILWSKGTEI  LGNSTRVTVT  SDGTLIIRNI  SRSDEGKYTC  FAENFMGKAN
                                      *
 501  STGILSVRDA  TKITLAPSSA  DINVGDNLTL  QCHASHDPTM  DLTFTWTLDD
 551  FPIDFDKPGG  HYRRASAKET  IGDLTILNAH  VRHGGKYTCM  AQTVVDGTSK
 601  EATVLVRGPP  GPPGGVVVRD  IGDTTVQLSW  SRGFDNHSPI  AKYTLQARTP
 651  PSGKWKQVRT  NPVNIEGNAE  TAQVLGLMPW  MDYEFRVSAS  NILGTGEPSG
 701  PSSKIRTKEA  VPSVAPSGLS  GGGGAPGELI  INWTPVSREY  QNGDGFGYLL
                                      *
 751  SFRRQGSSSW  QTARVPGADA  QYFVYGNDSI  QPYTPFEVKI  RSYNRRGDGP
                                                  *
 801  ESLTALVYSA  EEEPRVAPAK  VWAKGSSSSE  MNVSWEPVLQ  DMNGILLGYE
 851  IRYWKAGDNE  AAADRVRTAG  LDTSARVTGL  NPNTKYHVTV  RAYNRAGTGP
                    *                                        *
 901  ASPSADAMTV  KPPPRRPPGN  ISWTFSSSSL  SLKWDPVVPL  RNESTVTGYK
 951  MLYQNDLHPT  PTLHLTSKNW  IEIPVPEDIG  HALVQIRTTG  PGGDGIPAEV
1001  HIVRNGGTSM  MVESAAARPA  HPGPAFSCMV  ILMLAGYQKL
```

TAG-1

Domain structure
1–30	Signal sequence
56–117	Ig-1
150–213	Ig-2
256–312	Ig-3
343–401	Ig-4
435–494	Ig-5
525–593	Ig-6
629–694	FN-1
733–800	FN-2
835–914	FN-3
934–995	FN-4

Ala 1015 is the potential attachment site for the GPI-anchor.

Database accession numbers
	PIR	SWISSPROT	EMBL/GENBANK	REFERENCE
Rat	A34695	P22063	M31725	
Chick	S22128	P28685	X63101	4

Alternative forms

A soluble form of TAG-1 is detected in the supernatant of cultured neurons and is thought to be produced by alternative post-translational processing [3].

References
[1] Dodd, J. et al. (1988) Neuron 1, 105–116.
[2] **Furley, A.J. et al. (1990) Cell 61, 157–170.**
[3] Karagogeos, D. et al. (1991) Development 112, 51–67.
[4] Zuellig, R.A. et al. (1992) Eur. J. Biochem. 204, 453–463.
[5] Stoeckli, E.T. et al. (1991) J. Cell Biol. 112, 449–455.
[6] Kuhn, T.B. et al. (1992) J. Cell Biol. 115, 1113–1126.

VCAM-1 (vascular cell adhesion molecule) INCAM-110

Family
Immunoglobulin superfamily, C2 subset.

Cellular distribution
VCAM-1 is present on activated endothelial cells, tissue macrophages, dendritic cells, bone marrow fibroblasts, myoblasts and myotubes.

Function
VCAM-1 is important in the recruitment of leukocytes to sites of inflammation. VCAM-1 mediates the adhesion of lymphocytes, monocytes and eosinophils to activated endothelium [reviewed in ref.1]. VCAM-1 is involved in the interaction of lymphocytes and dendritic cells and may, therefore, have a role in normal immune function [2]. VCAM-1 mediates melanoma adhesion to endothelial cells and may be involved in metastasis [3]. The binding of T cells to VCAM-1 can induce antigen receptor-dependent activation [4]. VCAM-1 may be involved in myogenesis [5].

Regulation of expression
VCAM-1 expression on endothelial cells is induced by IL1β, IL4, TNFα, IFNγ [6]. VCAM-1 expression is developmentally regulated during myogenesis [5].

Ligands
Integrin α4/β1 (VLA-4) [7], integrin α4/β7 [8].

Gene structure [10]

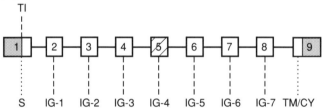

The VCAM-1 gene has potential NF-κB and AP1 binding sites.

Gene location and size
1P 31–32; 25 kb.

Structure [9]

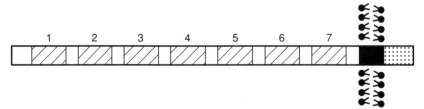

The pairs of domains 1 and 4, 2 and 5, 3 and 6 show the greatest homology, suggesting that the unit 1, 2, 3 may have been duplicated at some stage. There is evidence for the localization of ligand binding sites to domains 1 and 4 [8].

VCAM-1

Molecular weights
Polypeptide   68 659 and 78 744
SDS PAGE   90 000 and 110 000

Amino acid sequence (7-domain form, from human umbilical vein endothelial cells) [10].

```
  1  MPGKMVVILG ASNILWIMFA ASQAFKIETT PESRYLAQIG DSVSLTCSTT
 51  GCESPFFSWR TQIDSPLNGK VTNEGTTSTL TMNPVSFGNE HSYLCTATCE
101  SRKLEKGIQV EIYSFPKDPE IHLSGPLEAG KPITVKCSVA DVYPFDRLEI
151  DLLKGDHLMK SQEFLEDADR KSLETKSLEV TFTPVIEDIG KVLVCRAKLH
201  IDEMDSVPTV RQAVKELQVY ISPKNTVISV NPSTKLQEGG SVTMTCSSEG
251  LPAPEIFWSK KLDNGNLQHL SGNATLTLIA MRMEDSGIYV CEGVNLIGKN
301  RKEVELIVQE KPFTVEISPG PRIAAQIGDS VMLTCSVMGC ESPSFSWRTQ
351  IDSPLSGKVR SEGTNSTLTL SPVSFENEHS YLCTVTCGHK KLEKGIQVEL
401  YSFPRDPEIE MSGGLVNGSS VTVSCKVPSV YPLDRLEIEL LKGETILENI
451  EFLEDTDMKS LENKSLEMTF IPTIEDTGKA LVCQAKLHID DMEFEPKQRQ
501  STQTLYVNVA PRDTTVLVSP SSILEEGSSV NMTCLSQGFP APKILWSRQL
551  PNGELQPLSE NATLTLISTK MEDSGVYLCE GINQAGRSRK EVELIIQVTP
601  KDIKLTAFPS ESVKEGDTVI ISCTCGNVPE TWIILKKKAE TGDTVLKSID
651  GAYTIRKAQL KDAGVYECES KNKVGSQLRS LTLDVQGREN NKDYFSPELL
701  VLYFASSLII PAIGMIIYFA RKANMKGSYS LVEAQKSKV
```

Domain structure
  1–24   Signal sequence
 39–99   Ig-1
129–199   Ig-2
239–295   Ig-3
328–387   Ig-4
418–487   Ig-5
527–583   Ig-6
616–672   Ig-7
698–720   Transmembrane domain

Six potential N-linked glycosylation sites.

VCAM-1

Amino acid sequence (6-domain form, from human umbilical vein endothelial cells) [11]

```
  1  MPGKMVVILG ASNILWIMFA ASQAFKIETT PESRVLAQIG DSVSLTCSTT
                                ↓
 51  GCESPFFSWR TQIDSPLNGK VTNEGTTSTL TMNPVSFGNE HSYLCTATCE

101  SRKLEKGIQV EIYSFPKDPE IHLSGPLEAG KPITVKCSVA DVYPFDRLEI

151  DLLKGDHLMK SQEFLEDADR KSLETKSLEV TFTPVIEDIG KVLVCRAKLH

201  IDEMDSVPTV RQAVKELQVY ISPKNTVISV NPSTKLQEGG SVTMTCSSEG
                                    *
251  LPAPEIFWSK KLDNGNLQHL SGNATLTLIA MRMEDSGIYV CEGVNLIGKN
                                    *
301  RKEVELIVQA FPRDPEIEMS GGLVNGSSVT VSCKVPSVYP LDRLEIELLK
                           *
351  GETILENIEF LEDTDMKSLE NKSLEMTFIP TIEDTGKALV CQAKLHIDDM
                                                  *
401  EFEPKQRQST QTLYVNVAPR DTTVLVSPSS ILEEGSSVNM TCLSQGFPAP
                     *
451  KILWSRQLPN GELQPLSENA TLTLISTKME DSGVYLCEGI NQAGRSRKEV

501  ELIIQVTPKD IKLTAFPSES VKEGDTVIIS CTCGNVPETW IILKKKAETG

551  DTVLKSIDGA YTIRKAQLKD AGVYECESKN KVGSQLRSLT LDVQGRENNK

601  DYFSPELLVL YFASSLIIPA IGMIIYFARK ANMKGSYSLV EAQKSKV
```

Domain structure

1–24	Signal sequence
43–99	Ig-1
129–199	Ig-2
239–295	Ig-3
326–395	Ig-4
435–491	Ig-5
524–580	Ig-6
607–629	Transmembrane domain

Five potential N-linked glycosylation sites.

Database accession numbers

	PIR	SWISSPROT	EMBL/GENBANK
Human 7-domain		P19320	X53051
Human 6-domain	A33758		

Alternative forms

A variant lacking the fourth Ig-like domain is produced by alternative splicing [12].

References

1. Lobb, R. et al. (1991) In Cellular and Molecular Mechanisms of Inflammation: Vascular Adhesion Molecules (Cochrane, C.G. and Gimbrone, M.A. (eds) Academic Press, London, pp. 151–167.
2. Koopman, G. et al. (1991) J. Exp. Med. 173, 1297–1304.
3. Taichman, D.B. et al. (1991) Cell Regulation 2, 47–55.
4. Damle, N.K. and Aruffo, A. (1991) Proc. Natl Acad. Sci. USA 88, 6403–6407.
5. Rosen, G.D. et al. (1992) Cell 69, 1107–1119.
6. Masinovsky, B. et al. (1990) J. Immunol. 145, 2886–2895
7. Elices, M.J. et al. (1990) Cell 60, 577–584.
8. Ruegg, C. et al. (1992) J. Cell Biol. 117, 179–189.
9. Polte, T. et al. (1991) DNA Cell Biol. 10, 349–357.
10. Polte, T. et al. (1990) Nucleic Acids Res. 18, 5901–5901.
11. Osborn, L. et al. (1989) Cell 59, 1203–1211.
12. Cybulsky, M.I. et al. (1991) Proc. Natl Acad. Sci. USA 88, 7859–7863.

Vitronectin receptor

CD51/CD61, integrin αVβ3.

Family
β3 integrin (dimer of αV and β3 subunits).

Cellular distribution
Endothelial cells, some B cells, monocytes/macrophages, platelets and tumour cells.

Function
Adhesion to vitronectin [1,2]. Also binds vWF, fibrinogen and thrombospondin. Mediates platelet aggregation and endothelial cell adhesion to vitronectin and other extracellular matrix proteins. αVβ3 also functions as a laminin receptor on microvascular endothelial cells [3]. May be involved in monocyte or macrophage migration through subendothelial matrices to sites of inflammation. A role for αVβ3 in melanoma cell invasion has recently been proposed [4].

αVβ3 has been identified as an accessory co-stimulator of γδT cells for IL4 production [5].

Regulation of expression
Upregulated on various cell lines by phorbol esters.

Ligands
Vitronectin (binds through an RGD sequence). Also binds fibrinogen, vWF, thrombospondin, fibronectin, osteopontin and collagen [6].

Gene structure
Unknown for αV. (See GPIIb-IIIa for β3 subunit.)

Gene location
Chromosome 2 (αV) [7].

Structure

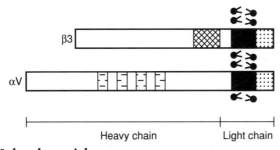

Molecular weights
Polypeptide   112 714 (αV subunit); 87 213 (β3 subunit); 85 647 (β5 subunit); 83 748 (β6 subunit); 81 168 (β8 subunit)
SDS PAGE   165 000 cleaved into 125 000/24 000 (αV); 105 000 (β3); 100 000 (β5); 106 000 (β6); 95 000 (β8)

Vitronectin receptor

Amino acid sequence (αV) (from IMR-90 human fibroblast cell line) [13]

```
   1  MAFPPRRRLR  LGPRGLPLLL  SGLLLPLCRA  FNLDVDSPAE  YSGPEGSYFG
  51  FAVDFFVPSA  SSRMFLLVGA  PKANTTQPGI  VEGGQVLKCD  WSSTRRCQPI
 101  EFDATGNRDY  AKDDPLEFKS  HQWFGASVRS  KQDKILACAP  LYHWRTEMKQ
 151  EREPVGTCFL  QDGTKTVEYA  PCRSQDIDAD  GQGFCQGGFS  IDFTKADRVL
 201  LGGPGSFYWQ  GQLISDQVAE  IVSKYDPNVY  SIKYNNQLAT  RTAQAIFDDS
 251  YLGYSVAVGD  FNGDGIDDFV  SGVPRAARTL  GMVYIYDGKN  MSSLYNFTGE
 301  QMAAYFGFSV  AATDINGDDY  ADVFIGAPLF  MDRGSDGKLQ  EVGQVSVSLQ
 351  RASGDFQTTK  LNGFEVFARF  GSAIAPLGDL  DQDGFNDIAI  AAPYGGEDKK
 401  GIVYIFNGRS  TGLNAVPSQI  LEGQWAARSM  PPSFGYSMKG  ATDIDKNGYP
 451  DLIVGAFGVD  RAILYRARPV  ITVNAGLEVY  PSILNQDNKT  CSLPGTALKV
 501  SCFNVRFCLK  ADGKGVLPRK  LNFQVELLLD  KLKQKGAIRR  ALFLYSRSPS
 551  HSKNMTISRG  GLMQCEELIA  YLRDESEFRD  KLTPITIFME  YRLDYRTAAD
 601  TTGLQPILNQ  FTPANISRQA  HILLDCGEDN  VCKPKLEVSV  DSDQKKIYIG
 651  DDNPLTLIVK  AQNQGEGAYE  AELIVSIPLQ  ADFIGVVRNN  EALARLSCAF
 701  KTENQTRQVV  CDLGNPMKAG  TQLLAGLRFS  VHQQSEMDTS  VKFDLQIQSS
 751  NLFDKVSPVV  SHKVDLAVLA  AVEIRGVSSP  DHIFLPIPNW  EHKENPETEE
 801  DVGPVVQHIY  ELRNNGPSSF  SKAMLHLQWP  YKYNNNTLLY  ILHYDIDGPM
 851  NCTSDMEINP  LRIKISSLQT  TEKNDTVAGQ  GERDHLITKR  DLALSEGDIH
 901  TLGCGVAQCL  KIVCQVGRLD  RGKSAILYVK  SLLWTETFMN  KENQNHSYSL
 951  KSSASFNVIE  FPYKNLPIED  ITNSTLVTTN  VTWGIQPAPM  PVPVWVIILA
1001  VLAGLLLLAV  LVFVMYRMGF  FKRVRPPQEE  QEREQLQPHE  NGEGNSET
```

Vitronectin receptor

Domain structure
The αV chain is composed of heavy and light chains linked by a disulphide bond. Four divalent cation binding sites in αV subunit.
1–30   Signal sequence
31–890 Heavy chain
891–1048 Light chain
993–1016 Transmembrane domain
1017–1048 Cytoplasmic domain
Thirteen potential N-linked glycosylation sites.

Amino acid sequence (β1)
See entry for VLA-1.

Amino acid sequence (β3)
See entry for GPIIb–IIIa.

Amino acid sequence (β5) (from human thymic epithelial cells) [9]

```
  1  MPRAPAPLYA CLLGLCALLP RLAGLNICTS GSATSCEECL LIHPKCAWCS
 51  KEDFGSPRSI TSRCDLRANL VKNGCGGEIE SPASSFHVLR SLPLSSKGSG
101  SAGWDVIQMT PQEIAVNLRP GDKTTFQLQV RQVEDYPVDL YYLMDLSLSM
151  KDDLDNIRSL GTKLAEEMRK LTSNFRLGFG SFVDKDISPF SYTAPRYQTN
201  PCIGYKLFPN CVPSFGFRHL LPLTDRVDSF NEEVRKQRVS RNRDAPEGGF
251  DAVLQAAVCK EKIGWRKDAL HLLVFTTDDV PHIALDGKLG GLVQPHDGQC
301  HLNEANEYTA SNQMDYPSLA LLGEKLAENN INLIFAVTKN HYMLYKNFTA
351  LIPGTTVEIL DGDSKNIIQL IINAYNSIRS KVELSVWDQP EDLNLFFTAT
401  CQDGVSYPGQ RKCEGLKIGD TASFEVSLEA RSCPSRHTEH VFALRPVGFR
451  DSLEVGVTYN CTCGCSVGLE PNSARCNGSG TYVCGLCECS PGYLGTRCEC
501  QDGENQSVYQ NLCREAEGKP LCSGRGDCSC NQCSCFESEF GKIYGPFCEC
551  DNFSCARNKG VLCSGHGECH CGECKCHAGY IGDNCNCSTD ISTCRGRDGQ
601  ICSERGHCLC GQCQCTEPGA FGEMCEKCPT CPDACSTKRD CVECLLLHSG
651  KPDNQTCHSL CRDEVITWVD TIVKDDQEAV LCFYKTAKDC VMMFTYVELP
701  SGKSNLTVLR EPECGNTPNA MTILLAVVGS ILLVGLALLA IWKLLVTIHD
751  RREFAKFQSE RSRARYEMAS NPLYRKPIST HTVDFTFNKF NKSYNGTVD
```

Vitronectin receptor

Domain structure
1–23	Signal sequence
24–719	Extracellular domain
464–512	Cysteine-rich repeat
513–554	Cysteine-rich repeat
555–593	Cysteine-rich repeat
594–630	Cysteine-rich repeat
720–742	Transmembrane domain
743–799	Cytoplasmic domain

Ten potential N-linked glycosylation sites.

Amino acid sequence (β6) (from the human pancreatic carcinoma cell line FG-2) [14]

```
                                            ↓                              *
  1  MGIELLCLFF LFLGRNDSRT RWLCLGGAET CEDCLLIGPQ CAWCAQENFT
                                                           *
 51  HPSGVGERCD TPANLLAKGC QLNFIENPVS QVEILKNKPL SVGRQKNSSD

101  IVQIAPQSLI LKLRPGGAQT LQVHVRQTED YPVDLYYLMD LSASMDDDLN

151  TIKELGSGLS KEMSKLTSNF RLGFGSFVEK PVSPFVKTTP EEIANPCSSI

201  PYFCLPTFGF KHILPLTNDA ERFNEIVKNQ KISANIDTPE GGFDAIMQAA
              *
251  VCKEKIGWRN DSLHLLVFVS DADSHFGMDS KLAGIVIPND GLCHLDSKNE

301  YSMSTVLEYP TIGQLIDKLV QNNVLLIFAV TQEQVHLYEN YAKLIPGATV
                                                *          *
351  GLLQKDSGNI LQLIISAYEE LRSEVELEVL GDTEGLNLSF TAICNNGTLF

401  QHQKKCSHMK VGDTASFSVT VNIPHCERRS RHIIIKPVGL GDALELLVSP
                    *          *
451  ECNCDCQKEV EVNSSKCHHG NGSFQCGVCA CHPGHMGPRC ECGEDMLSTD
                                                          *
501  SCKEAPDHPS CSGRGDCYCG QCICHLSPYG NIYGPYCQCD NFSCVRHKGL
                                  *
551  LCGGNGDCDC GECVCRSGWT GEYCNCTTST DSCVSEDGVL CSGRGDCVCG

601  KCVCTNPGAS GPTCERCPTC GDPCNSKRSC IECHLSAAGQ AGEECVDKCK

651  LAGATISEEE DFSKDGSVSC SLQGENECLI TFLITTDNEG KTIIHSINEK

701  DCPKPPNIPM IMLGVSLATL LIGVVLLCIW KLLVSFHDRK EVAKFEAERS

751  KAKWQTGTNP LYRGSTSTFK NVTYKHREKQ KVDLSTDC
```

Vitronectin receptor

Domain structure
1–19	Signal sequence
20–707	Extracellular domain
456–501	Cysteine-rich repeat
502–543	Cysteine-rich repeat
544–582	Cysteine-rich repeat
583–619	Cysteine-rich repeat
708–736	Transmembrane domain
737–788	Cytoplasmic domain

Nine potential N-linked glycosylation sites.

Amino acid sequence (β8) (from human placenta and MG63 osteosarcoma cells) [12]

```
  1  MCGSALAFFT AAFVCLQNDR RGPASFLWAA WVFSLVLGLG QGEDNRCASS
 51  NAASCARCLA LGPECGWCVQ EDFISGGSRS ERCDIVSNLI SKGCSVDSIE
101  YPSVHVIIPT ENEINTQVTP GEVSIQLRPG AEANFMLKVH PLKKYPVDLY
151  YLVDVSASMH NNIEKLNSVG NDLSRKMAFF SRDFRLGFGS YVDKTVSPYI
201  SIHPERIHNQ CSDYNLDCMP PHGYIHVLSL TENITEFEKA VHRQKISGNI
251  DTPEGGFDAM LQAAVCESHI GWRKEAKRLL LVMTDQTSHL ALDSKLAGIV
301  VPNDGNCHLK NNVYVKSTTM EHPSLGQLSE KLIDNNINVI FAVQGKQFHW
351  YKDLLPLLPG TIAGEIESKA ANLNNLVVEA YQKLISEVKV QVENQVQGIY
401  FNITAICPDG SRKPGMEGCR NVTSNDEVLF NVTVTMKKCD VTGGKNYAII
451  KPIGFNETAK IHIHRNCSCQ CEDNRGPKGK CVDETFLDSK CFQCDENKCH
501  FDEDQFSSES CKSHKDQPVC SGRGVCVCGK CSCHKIKLGK VYGKYCEKDD
551  FSCPYHHGNL CAGHGECEAG RCQCFSGWEG DRCQCPSAAA QHCVNSKGQV
601  CSGRGTCVCG RCECTDPRSI GRFCEHCPTC YTACKENWNC MQCLHPHNLS
651  QAILDQCKTS CALMEQQHYV DQTSECFSSP SYLRIFFIIF IVTFLIGLLK
701  VLIIRQVILQ WNSNKIKSSS DYRVSASKKD KLILQSVCTR AVTYRREKPE
751  EIKMDISKLN AHETFRCNF
```

Domain structure

1–42	Signal sequence
43–681	Extracellular domain
471–510	Cysteine-rich repeat
511–552	Cysteine-rich repeat
553–592	Cysteine-rich repeat
593–629	Cysteine-rich repeat
682–710	Transmembrane domain
711–769	Cytoplasmic domain

Seven potential N-linked glycosylation sites.

Database accession numbers

	PIR	SWISSPROT	EMBL/GENBANK	REFERENCE
Human αV	A27421		M14648	
		P06756		13
Human β5	S03436		X53002	
			X13366	
		P10804		9
Human β6		P18564	M35198	
			J05522	
Human β8			M73780	
Rabbit β8			M73781	

Alternative forms

αV associates with the novel integrin subunit β5, forming an alternative vitronectin receptor in carcinoma, hepatoma cells and fibroblasts [8,9]. The β5 (also known as βx) form is not found on lymphoblastoid cells or platelets. The physiological role of αVβ5 is not yet known.

A third variant αVβ1 has been identified as a fibronectin receptor [10].

A fourth variant αVβ6 has been discovered on the surface of two human carcinoma cell lines (FG-2 and UCLA-P3). This form also binds fibronectin but not collagen I or vitronectin and appears to be RGD-dependent [11].

A recently cloned, novel integrin subunit β8, has been shown to associate with αV *in vitro* [12].

References
[1] Pytela, R. et al. (1985) Proc. Natl Acad. Sci. USA 82, 5766–5770.
[2] Cheresh, D.A. (1987) Proc. Natl Acad. Sci. USA 84, 6471–6475.
[3] Kramer, R.H. et al. (1990) J. Cell Biol. 111, 1233–1243.
[4] Seftor, R.E. et al. (1992) Proc. Natl Acad. Sci. USA 89, 1557–1561.
[5] Roberts, K. et al. (1991) J. Exp. Med. 173, 231–240.
[6] Hynes, R.O. (1992) Cell 69, 11–25.
[7] Sosnoski, D.M. et al. (1988) J. Clin. Invest. 81, 1993–1998.
[8] Cheresh, D.A. et al. (1989) Cell 57, 59–69.
[9] Ramaswamy, H. and Hemler, M.E. (1990) EMBO J. 9, 1561–1568.
[10] Vogel, B.E. et al. (1990) J. Biol. Chem. 265, 5934–5937.

[11] Busk, M. et al. (1992) J. Biol. Chem. 267, 5790–5796.
[12] Moyle, M. et al. (1991) J. Biol. Chem. 266, 19650–19658.
[13] Suzuki, S. (1987) J. Biol. Chem. 262, 14080–14085.
[14] Sheppard, D. et al. (1990) J. Biol. Chem. 265, 11502–11507.

VLA-1 CD49a/CD29, integrin α1β1

Family
β1 integrin (dimer of α1 and β1 subunits).

Cellular distribution
Activated T lymphocytes, monocytes, neuronal cells in culture [1], melanoma cells [2] and smooth muscle cells [3].

Function
Collagen and laminin binding [2,3]. May have a role in cell differentiation. VLA-1 is upregulated in a number of inflammatory diseases of human intestine but its role in the disease process is unclear [4].

Regulation of expression
Appears on T lymphocytes 2–4 weeks after activation with alloantigen or mitogen. Differentiation of SY5Y and IMR32 neuroblastoma cell lines induced by retinoic acid results in upregulation of VLA-1 [5].

Ligands
Collagen [2,3]. Laminin (E1 fragment – does not appear to involve an RGD-like sequence) [6]. Rat VLA-1 appears to bind laminin at two different sites (fragment P1 and E8) [7].

Gene structure
Several exons of the β1 gene have been located spanning at least 30 kb of genomic DNA [8]. From the seven exons so far identified, it appears that the exon/intron boundaries are clearly conserved between the β1 and β3 genes (β1 and β3 are thought to have a common evolutionary origin). A 5 kb fragment contains exons 1 and 2 corresponding to the 3' end of exon C, exon D and exon E of β3. A 4.5 kb fragment contains exons 3–6 and corresponds to the 3' end of exon I, exon J, exon K, exon L and exon M of β3 gene.

Gene location
5 (human α1 chain); 10p11.2 (human β1 chain) [9].

Structure

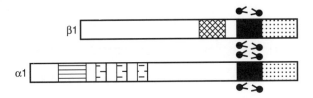

Molecular weights
Polypeptide   127 752 (rat α1); 86 240 (human β1)
SDS-PAGE   200 000 (rat α1); 210 000 (human α1); 130 000 (human β1)

VLA-1

Amino acid sequence (α1) (from rat pheochromocytoma PC12 cells) [10]

```
   1   MVPRRPASLE  VTVACIWLLT  VILGFCVSFN  VDVKNSMSFS  GPVEDMFGYT
  51   VQQYENEEGK  WVLIGSPLVG  QPKARTGDVY  KCPVGRERAM  PCVKLDLPVN
 101   TSIPNVTEIK  ENMTFGSTLV  TNPNGGFLAC  GPLYAYRCGH  LHYTTGICSD
 151   VSPTFQVVNS  FAPVQECSTQ  LDIVIVLDGS  NSIYPWESVI  AFLNDLLKRM
 201   DIGPKQTQVG  IVQYGENVTH  EFNLNKYSST  EEVLVAANKI  GRQGGLQTMT
 251   ALGIDTARKE  AFTEARGARR  GVKKVMVIVT  DGESHDNYRL  KQVIQDCEDE
 301   NIQRFSIAIL  GHYNRGNLST  EKFVEEIKSI  ASEPTEKHFF  NVSDELALVT
 351   IVKALGERIF  ALEATADQSA  ASFEMEMSQT  GFSAHYSQDW  VMLGAVGAYD
 401   WNGTVVMQKA  NQMVIPHNTT  FQTEPAKMNE  PLASYLGYTV  NSATIPGDVL
 451   YIAGQPRYNH  TGQVVIYKME  DGNINILQTL  GGEQIGSYFG  SVLTTIDIDK
 501   DSYTDLLVG   APMYMGTEKE  EQGKVYVYAV  NQTRFEYQMS  LEPIRQTCCS
 551   SLKDNSCTKE  NKNEPCGARF  GTAIAAVKDL  NVDGFNDVVI  GAPLEDDHAG
 601   AVYIYHGSGK  TIREAYAQRI  PSGGDGKTLK  FFGQSIHGEM  DLNGDGLTDV
 651   TIGGLGGAAL  FWARDVAVVK  VTMNFEPNKV  NIQKKNCRVE  GKETVCINAT
 701   MCFHVKLKSK  EDSIYEADLQ  YRVTLDSLRQ  ISRSFFSGTQ  ERKIQRNITV
 751   RESECIRHSF  YMLDKHDFQD  SVRVTLDFNL  TDPENGPVLD  DALPNSVHEH
 801   IPFAKDCGNK  ERCISDLTLN  VSTTEKSLLI  VKSQHDKFNV  SLTVKNKGDS
 851   AYNTRTVVQH  SPNLIFSGIE  EIQKDSCESN  QNITCRVGYP  FLRAGETVTF
 901   KIIFQFNTSH  LSENAIIHLS  ATSDSEEPLE  SLNDNEVNIS  IPVKYEVGLQ
 951   FYSSASEHHI  SVAANETIPE  FINSTEDIGN  EINVFYTIRK  RGHFPMPELQ
1001   LSISFPNLTA  DGYPVLYPIG  WSSSDNVNCR  PRSLEDPFGI  NSGKKMTISK
1051   SEVLKRGTIQ  DCSSTCGVAT  ITCSLLPSDL  SQVNVSLLLW  KPTFIRAHFS
1101   SLNLTLRGEL  KSENSSLTLS  SSNRKRELAI  QISKDGLPGR  VPLWVILLSA
1151   FAGLLLLMLL  ILALWKIGFF  KRPLKKKMEK
```

VLA-1

Domain structure
Seven tandem repeat domains with an additional I-domain (203 amino acids) inserted between repeats II and III. Contains three divalent metal binding sites (Ca^{2+}/ Mg^{2+}); however, Mn^{2+} has been shown to increase ECM protein binding.

1–27	Signal sequence
35–82	Repeat I
92–143	Repeat II
375–419	Repeat III
431–478	Repeat IV
485–535	Repeat V
565–620	Repeat VI
627–674	Repeat VII
161–385	I-domain
1139–1162	Transmembrane domain
1163–1180	Cytoplasmic domain

Twenty-four potential N-linked glycosylation sites.

Amino acid sequence (β1) (from human endothelial cells) [13]

```
  1  MNLQPIFWIG LISSVCCVFA QTDENRCLKA NAKSCGECIQ AGPNCGWCTN
 51  STFLQEGMPT SARCDDLEAL KKKGCPPDDI ENPRGSKDIK KNKNVTNRSK
101  GTAEKLKPED IHQIQPQQLV LRLRSGEPQT FTLKFKRAED YPIDLYYLMD
151  LSYSMKDDLE NVKSLGTDLM NEMRRITSDF RIGFGSFVEK TVMPYISTTP
201  AKLRNPCTSE QNCTTPFSYK NVLSLTNKGE VFNELVGKQR ISGNLDSPEG
251  GFDAIMQVAV CGSLIGWRNV TRLLVFSTDA GFHFAGDGKL GGIVLPNDGQ
301  CHLENNMYTM SHYYDYPSIA HLVQKLSENN IQTIFAVTEE FQPVYKELKN
351  LIPKSAVGTL SANSSNVIQL IIDAYNSLSS EVILENGKLS EGVTISYKSY
401  CKNGVNGTGE NGRKCSNISI GDEVQFEISI TSNKCPKKDS DSFKIRPLGF
451  TEEVEVILQY ICECECQSEG IPESPKCHEG NGTFECGACR CNEGRVGRHC
501  ECSTDEVNSE DMDAYCRKEN SSEICSNNGE CVCGQCVCRK RDNTNEIYSG
551  KFCECDNFNC DRSNGLICGG NGVCKCRVCE CNPNYTGSAC DCSLDTSTCE
601  ASNGQICNGR GICECGVCKC TDPKFQGQTC EMCQTCLGVC AEHKECVQCR
651  AFNKGEKKDT CTQECSYFNI TKVESRDKLP QPVQPDPVSH CKEKDVDDCW
701  FYFTYSVNGN NEVMVHVVEN PECPTGPDII PIVAGVVAGI VLIGLALLLI
751  WKLLMIIHDR REFAKFEKEK MNAKWDTGEN PIYKSAVTTV VNPKYEGK
```

Domain structure

The β subunit contains four cysteine-rich repeats in extracellular domain.

1–20	Signal sequence
466–515	Cysteine-rich repeat
516–559	Cysteine-rich repeat
560–598	Cysteine-rich repeat
599–635	Cysteine-rich repeat
729–751	Transmembrane domain
752–798	Cytoplasmic domain

Twelve potential N-linked glycosylation sites.

Database accession numbers

	PIR	SWISSPROT	EMBL/GENBANK	REFERENCE
Human β1	B27079			
		P05556		10
Mouse β1	S01659		Y00769	
		P09055		11
Rat α1		P18614	X52140	
Chicken β1	A23947	P07228	M14049	
X. laevis β1		P12606	M20140	
			J03736	

Alternative forms

Two variant forms of the β1 subunit have now been identified.

β1$^{3'v}$ is expressed at low levels compared to β1 in a number of cell types including HUVECs, lymphomas, neuroblastoma and hepatoma cell lines [12]. It is nine amino acids shorter than β1 and contains a unique 12 amino acid sequence in the C-terminal of the cytoplasmic domain. The new form does not contain a consensus sequence for tyrosine phosphorylation. Its function is unknown but it is likely that it mediates a novel type of membrane–cytoskeletal interaction during cell–cell or cell–extracellular matrix adhesion.

Variant β1s is a minor form of β1 expressed in platelets and various haematopoietic cell lines [11]. It encodes a new cytoplasmic domain which is 27 amino acids longer than β1 and contains a unique 48 amino acid sequence. β1s appears to be generated by alternative splicing of a new exon designated E6a which occurs between exons E6 and E7 (previously described in ref. 8 and see this section). This form may play a role in platelet adhesion mechanisms though it is not yet known which α subunit it is associated with. Interestingly, the cytoplasmic domain contains an SH2 sequence motif which may mediate binding to tyrosine phosphorylated proteins.

References
1. Ignatius, M. and Reichardt, L.F. (1988) Neuron 1, 713–725.
2. Kramer, R.H. and Marks, N. (1989) J. Biol. Chem. 264, 4684–4688.
3. Belkin, V.M. et al. (1990) J. Cell Biol. 111, 2159–2170.
4. MacDonald, T.T. et al. (1990) J. Clin. Pathol. 43, 313–315.

5 Rossino, P. et al. (1991) Cell Regulation 2, 1021–1033.
6 Hall, D.E. et al. (1990) J. Cell Biol. 110, 2175–2184.
7 Forsberg, E. et al. (1990) J. Biol. Chem. 265, 6376–6381.
8 Lanza, F. et al. (1990) J. Biol. Chem. 265, 18098–18103.
9 Goodfellow, P.N. et al. (1989) Ann. Human Genet. 53, 15–22.
10 Ignatius, M.J. et al. (1990) J. Cell Biol. 111, 709–720.
11 Languino, R.L. and Ruoslahti, E. (1992) J. Biol. Chem. 267, 7116–7120.
12 Altruda, F. et al. (1990) Gene 95, 261–266.
13 Argraves. W.S. et al. (1987) J. Cell Biol. 105, 1183–1190.

VLA-2

Platelet glycoprotein Ia–IIa, integrin α2β1, CD49b/CD29, ECMRI, collagen receptor

Family
β1 integrin (dimer of α2 and β1 subunits).

Cellular distribution
Widespread – B and T cells, platelets, fibroblasts; melanoma cell lines [1] and endothelial cells [2].

Function
Mg^{2+}-dependent adhesion of platelets to collagen [3]. VLA-2 is also selectively used as a collagen receptor on several types of mononuclear leukocytes [4]. It is required for contraction of a type I collagen matrix [5] and is therefore proposed to play an important role in wound healing. VLA-2 is also involved in the reorganization of collagen matrices by highly aggressive human melanoma cells [6]. VLA-2 may act as a laminin receptor on certain cell types but not on platelets or fibroblasts [7]. It has recently been shown to be a receptor for echovirus-1 [8].

Regulation of expression
Can be upregulated on lymphocytes in response to mitogen or antigen stimulation, and on fibroblasts in response to serum. Also upregulated by TGFβ. Withdrawal of stimuli reduces expression, as also occurs when cells become quiescent. Increased expression of α2β1 may be associated with malignant transformation [9] and increased tumorigenicity [10].

Ligands
Collagen type I (in the α1(I)–CB3 fragment), II, III and IV [11]. DGEA-containing peptides can inhibit the Mg^{2+}-dependent binding of platelet α2β1 to collagen [11], Laminin [7].

Gene structure
Unknown for α2 subunit.

Gene location
5q23–31(α2) [12].

Structure

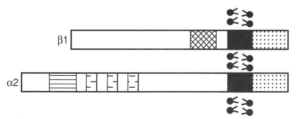

Molecular weights
Polypeptide   126 376 (α2); 86 240 (β1)
SDS-PAGE   155 000–165 000 (α2); 130 000 (β1)

VLA-2

Amino acid sequence (α2) (from human lung fibroblast IMR-90 cell line) [13]

```
   1  MGPERTGAAP LPLLLVLALS QGILNCCLAY NVGLPEAKIF SGPSSEQFGY
  51  AVQQFINPKG NWLLVGSPWS GFPENRMGDV YKCPVDLSTA TCEKLNLQTS
 101  TSIPNVTEMK TNMSLGLILT RNMGTGGFLT CGPLWAQQCG NQYYTTGVCS
 151  DISPDFQLSA SFSPATQPCP SLIDVVVVCD ESNSIYPWDA VKNFLEKFVQ
 201  GLDIGPTKTQ VGLIQYANNP RVVFNLNTYK TKEEMIVATS QTSQYGGDLT
 251  NTFGAIQYAR KYAYSAASGG RRSATKVMVV VTDGESHDGS MLKAVIDQCN
 301  HDNILRFGIA VLGYLNRNAL DTKNLIKEIK AIASIPTERY FFNVSDEAAL
 351  LEKAGTLGEQ IFSIEGTVQG GDNFQMEMSQ VGFSADYSSQ NDILMLGAVG
 401  AFGWSGTIVQ KTSHGHLIFP KQAFDQILQD RNHSSYLGYS VAAISTGEST
 451  HFVAGAPRAN YTGQIVLYSV NENGNITVIQ AHRGDQIGSY FGSVLCSVDV
 501  DKDTITDVLL VGAPMYMSDL KKEEGRVYLF TIKKGILGQH QFLEGPEGIE
 551  NTRFGSAIAA LSDINMDGFN DVIVGSPLEN QNSGAVYIYN GHQGTIRTKY
 601  SQKILGSDGA FRSHLQYFGR SLDGYGDLNG DSITDVSIGA FGQVVQLWSQ
 651  SIADVAIEAS FTPEKITLVN KNAQIILKLC FSAKFRPTKQ NNQVAIVYNI
 701  TLDADGFSSR VTSRGLFKEN NERCLQKNMV VNQAQSCPEH IIYIQEPSDV
 751  VNSLDLRVDI SLENPGTSPA LEAYSETAKV FSIPFHKDCG EDGLCISDLV
 801  LDVRQIPAAQ EQPFIVSNQN KRLTFSVTLK NKRESAYNTG IVVDFSENLF
 851  FASFSLPVDG TEVTCQVAAS QKSVACDVGY PALKREQQVT FTINFDFNLQ
 901  NLQNQASLSF QALSESQEEN KADNLVNLKI PLLYDAEIHL TRSTNINFYE
 951  ISSDGNVPSI VHSFEDVGPK FIFSLKVTTG SVPVSMATVI IHIPQYTKEK
1001  NPLMYLTGVQ TDKAGDISCN ADINPLKIGQ TSSSVSFKSE NFRHTKELNC
1051  RTASCSNVTC WLKDVHMKGE YFVNVTTRIW NGTFASSTFQ TVQLTAAAEI
1101  NTYNPEIYVI EDNTVTIPLM IMKPDEKAEV PTGVIIGSII AGILLLLALV
1151  AILWKLGFFK RKYEKMTKNP DEIDETTELS S
```

VLA-2

Domain structure
The α2 subunit contains seven homologous repeats of 28–41 amino acids each as well as an additional I-domain of 191 amino acids inserted between repeats II and III. Three metal ion binding sites in α2 subunit. Requires Mg^{2+} for binding to collagen [3].

1–29	Signal sequence
46–81	Repeat I
113–147	Repeat II
379–406	Repeat III
434–466	Repeat IV
489–528	Repeat V
552–588	Repeat VI
616–656	Repeat VII
188–378	I-domain
1133–1154	Transmembrane domain
1155–1181	Cytoplasmic domain

Ten potential N-linked glycosylation sites.

Amino acid sequence (β1)
See entry for VLA-1.

Database accession numbers

	PIR	SWISSPROT	EMBL/GENBANK	REFERENCE
Human α2	A33998		X17033	
		P17301		13

References
[1] Klein, C.E. et al. (1991) J. Invest. Dermatol. 96, 281–284.
[2] Giltay, J.C. et al. (1989) Blood 73, 1235–1241.
[3] Staatz, W.D. et al. (1989) J. Cell Biol. 108, 1917–1924.
[4] Goldman, R. et al. (1992) Eur. J. Immunol. 22, 1109–1114.
[5] Schiro, J.A. et al. (1991) Cell, 67, 403–410.
[6] Klein, C.E. et al. (1991) J. Cell. Biol. 115, 1427–1436.
[7] Elices, M.J. and Hemler, M.E. (1989) Proc. Natl Acad. Sci.USA 86, 9906–9910.
[8] Bergelson, J.M. et al. (1992) Science 255, 1718–1720.
[9] Chen, F. et al. (1991) J. Exp. Med. 173, 1111–1119.
[10] Chan, B.M.C. et al. (1991) Science 251, 1600–1602.
[11] Staatz., W.D. et al. (1991) J. Biol. Chem. 266, 7363–7367.
[12] Jaspers, M. et al. (1991) Somat. Cell. Mol. Genet. 17, 505–511.
[13] **Takada, Y. and Hemler, M.E. (1989) J. Cell Biol. 109, 397–407.**

VLA-3

ECMRII, integrin α3β1, CD49c/CD29

Other names
Laminin receptor, galactoprotein b3 (hamster) [13].

Family
β1 integrin (dimer of α3 and β1 chain).

Cellular distribution
Kidney glomerulus, thyroid, some basement membranes, B cells [1]. Nearly all culture cell lines except lymphoid.

Function
Extracellular matrix adhesion receptor which binds to multiple ligands involving both RGD and non-RGD sites [2,3]. May also be involved in cell–cell adhesion [4]. MAb-induced clustering of VLA-3 can result in phosphorylation of a 115–130 kD complex of proteins termed pp130 [5].

Regulation of expression
Induced by attachment of some cultured cells to extracellular matrix. In certain cell types VLA-3 is downregulated by TGFβ. VLA-3 expression also decreases when cells enter quiescence [1].

Ligands
Laminin (binds E3 fragment) [6]. Collagen (binding is best demonstrated in the absence of other collagen receptors) [2,7]. Fibronectin (RGD-dependent binding if VLA-5 is absent or depleted) [2]. Invasin [8] and epiligrin [9].

Gene structure
Unknown for α3 subunit.

Gene location
17 (α3) [10].

Structure

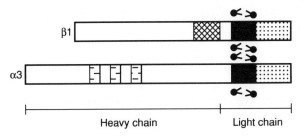

Molecular weights
Polypeptide   113 505 (α3)
SDS-PAGE   145 000–150 000 (α3); Cleaved into heavy and light chains of approximately 110 000 and 30 000 respectively.

VLA-3

Amino acid sequence (α3) (from endothelial cell λGT11 cDNA library) [11]

```
                                              ↓
   1  MGPGPSRAPR APRLMLCALA LMVAAGGCVV SAFNLDTRFL VVKEAGNPGS
                                            *
  51  LFGYSVALHR QTERQQRYLL LAGAPRELAV PDGYTNRTGA VYLCPLTAHK
          *
 101  DDCERMNITV KNDPGHHIIE DMWLGVTVAS QGPAGRVLVC AHRYTQVLWS
 151  GSEDQRRMVG KCYVRGNDLE LDSSDDWQTY HNEMCNSNTD YLETGMCQLG
 201  TSGGFTQNTV YFGAPGAYNW KGNSYMIQRK EWDLSEYSYK DPEDQGNLYI
                      *
 251  GYTMQVGSFI LHPKNITIVT GAPRHRHMGA VFLLSQEAGG DLRRRQVLEG
 301  SQVGAYFGSA IALADLNNDG WQDLLVGAPY YFERKEEVGG AIYVFMNQAG
 351  TSFPAHPSLL LHGPSGSAFG LSVASIGDIN QDGFQDIAVG APFEGLGKVY
 401  IYHSSSKGLL RQPQQVIHGE KLGLPGLATF GYSLSGQMDV DENFYPDLLV
                                                           *
 451  GSLSDHIVLL RARPVINIVH KTLVPRPAVL DPALCTATSC VQVELCFAYN
               *
 501  QSAGNPNYRR NITLAYTLEA DRDRRPPRLR FAGSESAVFH GFFSMPEMRC
                                 *
 551  QKLELLLMDN LRDKLRPIII SMNYSLPLRM PDRPRLGLRS LDAYPILNQA
        *
 601  QALENHTEVQ FQKECGPDNK CESNLQMRAA FVSEQQQKLS RLQYSRDVRK
        *                                                *
 651  LLLSINVTNT RTSERSGEDA HEALLTLVVP PALLLSSVRP PGACQANETI
 701  FCELGNPFKR NQRMELLIAF EVIGVTLHTR DLQVQLQLST SSHQDNLWPM
 751  ILTLLVDYTL QTSLSMVNHR LQSFFGGTVM GESGMKTVED VGSPLKYEFQ
                                                     *
 801  VGPMGEGLVG LGTLVLGLEW PYEVSNGKWL LYPTEITVHG NGSWPCRPPG
 851  DLINPLNLTL SDPGDRPSSP QRRRRQLDPG GGQGPPPVTL AAAKKAKSET
                                 *                *
 901  VLTCATGRAH CVWLECPIPD APVVTNVTVK ARVWNSTFIE DYRDFDRVRV
                             *
 951  NGWATLFLRT SIPTINMENK TTWFSVDIDS ELVEELPAEI ELWLVLVAVG
1001  AGLLLLGLII LLLWKCGFFK RARTRALYEA KRQKAEMKSQ PSETERLTDD
1051  Y
```

Domain structure

The α3 subunit contains seven homologous repeating domains and is cleaved into a heavy and light chain linked by a disulphide bond. Repeating domains V, VI and VII each contain a putative divalent metal cation binding site. Different ligands have different cation sensitivities. Fibronectin binding is inhibited by Ca^{2+}. Collagen type I and IV binding is not affected by Ca^{2+} [2,3].

1–32	Signal peptide
50–92	Repeat I
131–163	Repeat II
197–225	Repeat III
327–361	Repeat IV
384–422	Repeat V
447–479	Repeat VI
507–544	Repeat VII
992–1019	Transmembrane domain
1020–1051	Cytoplasmic domain

Thirteen potential N-linked glycosylation sites.

Amino acid sequence (β1)

See entry for VLA-1.

Database accession numbers

	PIR	SWISSPROT	EMBL/GENBANK	REFERENCE
Human α3			M59911	11
Hamster α3	A35761	P17852	J05281	

Alternative forms

A variant form of the α3 subunit (α3B), thought to arise by alternative splicing, has been detected in mouse brain and heart. α3B contains an alternative cytoplasmic domain [12].

References

[1] Hemler, M.E. (1990) Annu. Rev. Immunol. 8, 365–400.
[2] Elices, M.J. et al. (1991) J. Cell Biol. 112, 169–181.
[3] Hemler, M.E. et al. (1990) Cell. Differ. Dev. 32, 229–238.
[4] Kaufmann, R. et al. (1989) J. Cell Biol. 109, 1807–1815.
[5] Kornberg, L.J. et al. (1991) Proc. Natl Acad. Sci. USA 88, 8392–8396.
[6] Gehlsen, K.R. et al. (1989) J. Biol. Chem. 264, 19034–19038.
[7] **Hemler, M.E. (1991) In Receptors for Extracellular Matrix (MacDonald, J. and Mecham, R. eds) Academic Press, New York, pp. 255–300.**
[8] Isberg, R.R. and Leong, J.M. (1990) Cell 60, 861–871.
[9] Carter, W.G. et al. (1991) Cell 65, 599–610.
[10] Rettig, W.J. et al. (1984) Proc. Natl Acad. Sci. USA 81, 6437–6441.
[11] Takada, Y. et al. (1991) J. Cell Biol. 115, 257–266.
[12] Tamura, R.N et al. (1991) Proc. Natl Acad. Sci. USA 88, 10183–10187.
[13] Tsuji, T. et al. (1991) J. Biochem. (Tokyo) 109, 659–665.

 Very late antigen-4, CD49d/CD29, integrin α4β1, LPAM-2

Family
β1 integrin (dimer of α4 and β1 subunits)

Cellular distribution
Thymocytes, peripheral blood lymphocytes, monocytes, B and T cell lines, NK cells, myelomonocytic cells, eosinophils, erythroblastic precursor cells, many melanomas. Adherent cell lines generally show weak expression. Also detected on muscle cells in culture.

Function
The functions of VLA-4 have recently been reviewed [1].

VLA-4 is a receptor for fibronectin. Fibronectin binding by VLA-4 can act as a co-stimulus with cross-linking of the T cell receptor–CD3 complex in stimulating T cell proliferation [2]. VLA-4 also functions as a cell–cell adhesion molecule, promoting heterotypic adhesion between B and T cells and between helper and suppressor T cells. It also mediates homotypic adhesion between B and T lymphoblastoid cell lines. VLA-4 plays a major role in mediating mononuclear leukocyte migration during inflammation [1,3]. VLA-4 is also involved in the normal development of skeletal muscle [4].

It has recently been shown that VLA-4 binding on T cells stimulates tyrosine phosphorylation of an as yet uncharacterized 105 kD protein in resting peripheral blood T cells and in the T cell line H9 [5]. This phosphorylation can also be triggered by anti-α4 monoclonal antibodies.

Regulation of expression
Expression on T cells is slowly increased upon activation.

Ligands
VCAM-1 [6]; fibronectin (non-RGD-dependent site in CS-1 region (EILDV sequence); also binds to CS5 and Hep11 regions) [7,8].

Gene structure
Characterization of the α4 subunit gene promoter reveals the presence of a consensus binding site for the muscle-specific transcription factor Myo D, and an AP1 site, which may be important in the control of α4 gene expression [9].

Gene location
2q31–32 (α4)[10].

VLA-4

Structure

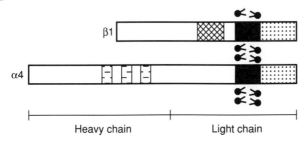

Molecular weights

Polypeptide 111 226 (α4); 86 240 (β1)

SDS-PAGE 150 000 (α4) which is cleaved to yield two fragments of 80 000 and 70 000. The 150 kD and 80,70 kD forms are both found in resting lymphocytes whereas only the 80,70 kD form is detectable after activation, possibly due to the existence of specific proteases [11,12].
130 000 (β1); 120 000 (β7)

Amino acid sequence (α4) (cDNA cloned from HPB-MLT cells) [13]

```
  1   MFPTESAWLG KRGANPGPEA AVRETVMLLL CLGVPTGRPY NVDTESALLY
 51   QGPHNTLFGY SVVLHSHGAN RWLLVGAPTA NWLANASVIN PGAIYRCRIG
101   KNPGQTCEQL QLGSPNGEPC GKTCLEERDN QWLGVTLSRQ PGENGSIVTC
151   GHRWKNIFYI KNENKLPTGG CYGVPPDLRT ELSKRIAPCY QDYVKKFGEN
201   FASCQAGISS FYTKDLIVMG APGSSYWTGS LFVYNITTNK YKAFLDKQNQ
251   VKFGSYLGYS VGAGHFRSQH TTEVVGGAPQ HEQIGKAYIF SIDEKELNIL
301   HEMKGKKLGS YFGASVCAVD LNADGFSDLL VGAPMQSTIR EEGRVFVYIN
351   SGSGAVMNAM ETNLVGSDKY AARFGESIVN LGDIDNDGFE DVAIGAPQED
401   DLQGAIYIYN GRADGISSTF SQRIEGLQIS KSLSMFGQSI SGQIDADNNG
451   YVDVAVGAFR SDSAVLLRTR PVVIVDASLS HPESVNRTKF DCVENGWPSV
501   CIDLTLCFSY KGKEVPGYIV LFYNMSLDVN RKAESPPRFY FSSNGTSDVI
551   TGSIQVSSRE ANCRTHQAFM RKDVRDILTP IQIEAAYHLG PHVISKRSTE
601   EFPPLQPILQ QKKEKDIMKK TINFARFCAH ENCSADLQVS AKIGFLKPHE
651   NKTYLAVGSM KTLMLNVSLF NAGDDAYETT LHVKLPVGLY FIKILELEEK
```

VLA-4

```
701   QINCEVTDNS GVVQLDCSIG YIYVDHLSRI DISFLLDVSS LSRAEEDLSI
751   TVHATCENEE EMDNLKHSRV TVAIPLKYEV KLTVHGFVNP TSFVYGSNDE
                      *                 *
801   NEPETCMVEK MNLTFHVINT GNSMAPNVSV EIMVPNSFSP QTDKLFNILD
851   VQTTTGECHF ENYQRVCALE QQKSAMQTLK GIVRFLSKTD KRLLYCIKAD
901   PHCLNFLCNF GKMESGKEAS VHIQLEGRPS ILEMDETSAL KFEIRATGFP
951   EPNPRVIELN KDENVAHVLL EGLHHQRPKR YFT<u>IVIISSS LLLGLIVLLL</u>
1001  <u>ISYVMWKAGF</u> FKRQYKSILQ EENRRDSWSY INSKSNDD
```

Domain structure

The α4 subunit contains seven homologous repeating domains spaced 22-37 amino acids apart. Cleavage of the α4 subunit is thought to occur between Arg_{597} and Ser_{598} [11].

1–39	Signal sequence
55–95	Domain I
131–172	Domain II
204–229	Domain III
255–288	Domain IV
310–346	Domain V
372–407	Domain VI
434–474	Domain VII
984–1006	Transmembrane domain
1007–1038	Cytoplasmic domain

Twelve potential N-linked glycosylation sites.

Three metal ion binding sites in α4 subunit domains V, VI and VII (Ca^{2+}-dependent).

Amino acid sequence (β1)

See entry for VLA-1.

Amino acid sequence (β7) (from SEA-activated human T lymphocytes) [14]

```
                         ↓
1    MVALPMVLVL LLVLSRGESE LDAKIPSTGD ATEWRNPHLS MLGSCQPAPS
                  *
51   CQKCILSHPS CAWCKQLNFT ASGEAEARRC ARREELLARG CPLEELEEPR
101  GQQEVLQDQP LSQGARGEGA TQLAPQRVRV TLRRGEPQQL QVRFLRAEGY
```

VLA-4

```
151 PVDLYYLMDL SSSMKDDLER VRQLGHALLV RLQEVTHSVR IGFGSFVDKT
201 VLPFVSTVPS KLRHPCPTRL ERCQSPFSFH HVLSLTGDAQ AFEREVGRQS
                                  *
251 VSGNLDSPEG GFDAILQAAL CQEQIGWRNV SRLLVFTSDD TFHTAGDGKL
301 GGIFMPSDGS CHLDSNGLYS RSTEFDYPSV GQVAQALSAA NIQPIFAVTS
351 AALPVYQELS KLIPKSAVGE LSEDSSNVVQ LIMDAYNSLS STVTLEHSSL
                                  *
401 PPGVHISYES QCEGPEKREG KAEDRGQCNH VRINQTVTFW VSLQATHCLP
                             *
451 EPHLLRLRAL GFSEELIVEL HTLCDCNCSD TQPQAPHCSD AGHLQCGVCS
                           *
501 CAPGALGRLC ECSVAELSSP DLESGCRAPN GTGPLCSGKG HCQCGRCSCS
                                  *
551 GQSSGHLCEC DDASCERHEG MLCGGFGRCQ CGVCHCHANR TGRACECSGD
601 MDSCISPEGG LCSGHGRCKC NRCQCLDGYY GALCDQCPGC KTPCERHRDC
         *           *
651 AECGAFRTGP LATNCSTACA HTNVTLALAP ILDDGWCKER TLDNQLFFFL
701 VEDDARATVV LRVRPQEKGA DHTQAIVLGW VGGIVAVGLG LVLAYRLSVE
751 IYDRREYSRF EKEQQQLNWK QDSNPLYKSA ITTTINPRFQ EADSPTL
```

Domain structure
  1–17   Signal sequence
  18–724 Extracellular domain
  725–745 Transmembrane domain
  746–797 Cytoplasmic domain
  478–525 Cysteine-rich repeat
  526–564 Cysteine-rich repeat
  565–603 Cysteine-rich repeat
  604–639 Cysteine-rich repeat
  8 potential N-linked glycosylation sites

Database accession numbers

	PIR	SWISSPROT	EMBL/GENBANK	REFERENCE
Human α4	S06046	P13612	X16983	
			X15356	
Human β7		M68892		14
			M62880	15
Mouse α4			X53176	

Alternative forms

VLA-4$_{alt}$ (LPAM-1 or murine Peyer's patch homing receptor) is a dimer of the α4 subunit and a novel β chain βp found on mouse lymphoid cells and normal mesenteric node lymphocytes. βp has now been shown to be identical to the novel integrin subunit β7 [16]. Its ligand on ppHEV is not yet known [17].

The α4 chain is also associated with the novel β7 integrin subunit in human mucosal lymphocytes [18]. β7 can be induced late by activation of peripheral blood T cells. The α4β7 complex appears to be minimally active as a receptor for VLA-4 ligands [19] but is probably the receptor for mucosal vascular addressin [20].

The β7 chain also associates with a novel α subunit designated α_{IEL} [18].

References
[1] **Hemler, M.E. (1990) Immunol. Rev. 114, 45–65.**
[2] Davis, L.S. et al. (1990) J. Immunol. 145, 785–793.
[3] Bednarczyk, J.L. and MacIntyre, B.W. (1990) J. Immunol. 144, 777–784.
[4] Rosen, G.D. et al. (1992) Cell 69, 1107–1119.
[5] Nojima, Y. et al. (1992) J. Exp. Med. 175, 1045–1053.
[6] Elices, M.J. et al. (1990) Cell 60, 577–584.
[7] Wayner, E.A. et al. (1989) J. Cell Biol. 109, 1321–1326.
[8] Mould, A.P. and Humphries, M.J. (1991) EMBO J. 10, 4089–4095.
[9] Rosen, G.D. et al. (1991) Proc. Natl Acad. Sci. USA 88, 4094–4098.
[10] Zhang, Z. et al. (1991) Blood 78, 2396–2399.
[11] Rubio, M. et al. (1992) Eur. J. Immunol. 22, 1099–1102.
[12] Teixido, J. et al. (1992) J. Biol. Chem. 267, 1786–1791.
[13] Takada, Y. et al. (1989) EMBO J. 8, 1361–1368.
[14] Yuan, J. et al. (1990) Int. Immunol. 2, 1097–1108.
[15] Erle, D.J. et al. (1991) J. Biol. Chem. 266, 11009–11016.
[16] Ruegg, C. et al. (1992) J. Cell Biol. 117, 179–189.
[17] Holzman, B. et al. (1989) Cell 56, 37–46.
[18] Parker, C.M. et al. (1992) Proc. Natl Acad. Sci. USA 89, 1924–1928.
[19] Chan, B.M.C. et al. (1992) J. Biol. Chem. 267, 8366–8370.
[20] Nakache, M. et al. (1989) Nature 337, 179–181.

VLA-5

CD49e/CD29, integrin αF, GPIc–IIa, fibronectin receptor

Family
β1 integrin (dimer of αF or α5 chain with β1).

Cellular distribution
Widespread – monocytes and monocytoid cell lines, leukocytes, memory T cells, fibroblasts, platelets [1,2] and muscle cells [3].

Function
VLA-5 functions as a fibronectin receptor and mediates binding of B and T lymphocytes to fibronectin. Fibronectin binding to VLA-5 on T cells acts a co-stimulus with the cross-linking of the T cell receptor–CD3 complex to stimulate T cell activation and proliferation. VLA-5 may also be involved in cell migration [4-7].

Fibronectin binding triggers phosphorylation of a 120 kD protein [8]; the β1 cytoplasmic domain is essential for this activity. Binding of fibronectin also results in AP1 transcription factor induction in CD4+ T cells [9] and in collagenase and stromelysin gene induction in fibroblasts [1]. Binding of fibronectin by VLA-5 activates an Na/H antiporter which may be important in regulating cell growth [10]. In contrast, during differentiation of F9 teratocarcinoma stem cells, although VLA-5 is expressed on the cell surface, it loses activity which correlates with a loss of phosphoserine labelling [11].

Regulation of expression
Constitutively expressed. However, on B cells, VLA-5 expression is regulated during B cell development, being expressed at a very early stage and then again after activation [12].

Ligands
Fibronectin (RGD-dependent) and invasin [13].

Gene structure
Unknown for α5 subunit.

Gene location
12q11–13 (α5) [14].

Structure

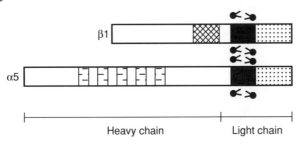

Heavy chain — Light chain

VLA-5

Molecular weights
Polypeptide   110 024 (α5)
SDS-PAGE   160 000 (α5). Normally appears cleaved as two fragments of MW 135 000 and 25 000

Amino acid sequence (α5) (from human placenta λGT11 cDNA library) [4]

```
   1  MGSRTPESPL HAVQLRWGPR RRPPLVPLLL LLVPPPPRVG G↓FNLDAEAPA
  51  VLSGPPGSFF GFSVEFYRPG TDGVSVLVGA PKANTSQPGV LQGGAVYLCP
 101  WGASPTQCTP IEFDSKGSRL LESSLSSSEG EEPVEYKSLQ WFGATVRAHG
 151  SSILACAPLY SWRTEKEPLS DPVGTCYLST DNFTRILEYA PCRSDFSWAA
 201  GQGYCQGGFS AEFTKTGRVV LGGPGSYFWQ GQILSATQEQ IAESYYPEYL
 251  INLVQGQLQT RQASSIYDDS YLGYSVAVGE FSGDDTEDFV AGVPKGNLTY
 301  GYVTILNGSD IRSLYNFSGE QMASYFGYAV AATDVNGDGL DDLLVGAPLL
 351  MDRTPDGRPQ EVGRVYVYLQ HPAGIEPTPT LTLTGHDEFG RFGSSLTPLG
 401  DLDQDGYNDV AIGAPFGGET QQGVVFVFPG GPGGLGSKPS QVLQPLWAAS
 451  HTPDFFGSAL RGGRDLDGNG YPDLIVGSFG VDKAVVYRGR PIVSASASLT
 501  IFPAMFNPEE RSCSLEGNPV ACINLSFCLN ASGKHVADSI GFTVELQLDW
 551  QKQKGGVRRA LFLASRQATL TQTLLIQNGA REDCREMKIY LRNESEFRDK
 601  LSPIHIALNF SLDPQAPVDS HGLRPALHYQ SKSRIEDKAQ ILLDCGEDNI
 651  CVPDLQLEVF GEQNHVYLGD KNALNLTFHA QNVGEGGAYE AELRVTAPPE
 701  AEYSGLVRHP GNFSSLSCDY FAVNQSRLLV CDLGNPMKAG ASLWGGLRFT
 751  VPHLRDTKKT IQFDFQILSK NLNNSQSDVV SFRLSVEAQA QVTLNGVSKP
 801  EAVLFPVSDW HPRDQPQKEE DLGPAVHHVY ELINQGPSSI SQGVLELSCP
 851  QALEGQQLLY VTRVTGLNCT TNHPINPKGL ELDPEGSLHH QQKREAPSRS
 901  SASSGPQILK CPEAECFRLR CELGPLHQQE SQSLQLHFRV WAKTFLQREH
 951  QPFSLQCEAV YKALKMPYRI LPRQLPQKER QVATAVQWTK AEGSYGVPLW
1001  IIILAILFGL LLLGLLIYIL YKLGFFKRSL PYGTAMEKAQ LKPPATSDA
```

VLA-5

Domain structure

The α5 subunit is cleaved into a heavy and light chain which are linked by a disulphide bond.

1–41	Signal sequence
42–894	Heavy chain
895–1049	Light chain
996–1021	Transmembrane domain
1022–1049	Cytoplasmic domain

Fourteen potential N-linked glycosylation sites.

Five potential metal ion binding sites (Ca^{2+}-dependent).

Database accession numbers

	PIR	SWISSPROT	EMBL/GENBANK	REFERENCE
Human α5	A24697		X06256	
		P08648		5
Mouse α5	PL0103	P11688		
Mouse, C-terminal 408 amino acids			X15203	15

Amino acid sequence (β1)

See entry for VLA-1.

References

1 Sanchez-Madrid, F. and Corbi, A.L. (1993) Seminars Cell Biol. 3, 199–210.
2 Hemler, M.E. (1990) Annu. Rev. Immunol. 8, 365–400.
3 Bozyczko, G. et al. (1989) Exp. Cell Res. 183, 72–91.
4 Argraves, W.S. et al. (1987) J. Cell Biol. 105, 1183–1190.
5 **Argraves, W.S. et al. (1986) J. Biol. Chem. 261, 12922–12924.**
6 Pytela, R. et al. (1985) Cell 40, 191–198.
7 Shimizu, Y. et al. (1990) J. Immunol. 145, 59–67.
8 Guan, J.L. et al. (1991) Cell Regulation 2, 951–964.
9 Yamada, A. et al. (1991) J. Immunol. 146, 53–56.
10 Schwartz, M.A. et al. (1991) Proc. Natl Acad. Sci. USA 88, 7849–7853.
11 Dahl, S.C. and Grabel, L.B. (1989) J. Cell Biol. 108, 183–190.
12 Ballard, L.L. et al. (1991) Clin. Exp. Immunol. 84, 336–346.
13 Isberg, R.R. and Leong, J.M. (1990) Cell 60, 861–871.
14 Sosnoski, D.M. et al. (1988) J. Clin. Invest. 81, 1993–1998.
15 Holers, V.M. et al. (1989) J. Exp. Med. 169, 1589–1605.

VLA-6

Integrin α6β1, IcIIa, CD49f/CD29, laminin receptor

Family
β1 integrin (dimer of α6 subunit and β1 subunit).

Cellular distribution
Widespread: platelets, monocytes, T lymphocytes, thymocytes.

Function
VLA-6 mediates the adhesion of platelets to laminin [1]. It may also have a key role in mediating the effects of laminin during embryogenesis. Important in the induction of polarity during the differentiation of embryonic kidney mesenchyme into epithelium [2]. PMA treatment of macrophages leads to phosphorylation of the α6 subunit which may be required for laminin binding [3]. Laminin binding by VLA-6 may also be regulated by a post-translational mechanism in some neurons [4].

Regulation of expression
Constitutively expressed.

Ligands
Laminin (E8 region) [5]; invasin [6].

Gene structure
Unknown for α6 subunit.

Gene location
α6 chromosome 2 [7].

Structure

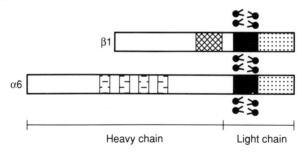

Molecular weights
Polypeptide   117 263 (α6)
SDS-PAGE   150 000. α6 is cleaved into heavy and light chains of MW 125 000 and 30 000.

VLA-6

Amino acid sequence (α6) (from a human pancreatic carcinoma cell line (FG))[8]

```
   1  MAAAGQLCLL  YLSAGLLSRL  GAAFNLDTRE  DNVIRKYGDP  GSLFGFSLAM

  51  HWQLQPEDKR  LLLVGAPRGE  ALPLQRANRT  GGLYSCDITA  RGPCTRIEFD

 101  NDADPTSESK  EDQWMGVTVQ  SQGPGGKVVT  CAHRYEKRQH  VNTKQESRDI

 151  FGRCYVLSQN  LRIEDDMDGG  DWSFCDGRLR  GHEKFGSCQQ  GVAATFTKDF

 201  HYIVFGAPGT  YNWKGIVRVE  QKNNTFFDMN  IFEDGPYEVG  GETEHDESLV

 251  PVPANSYLGF  SLDSGKGIVS  KDEITFVSGA  PRANHSGAVV  LLKRDMKSAH

 301  LLPEHIFDGE  GLASSFGYDV  AVVDLNKDGW  QDIVIGAPQY  FDRDGEVGGA

 351  VYVYMNQQGR  WNNVKPIRLN  GTKDSMFGIA  VKNIGDINQD  GYPDIAVGAP

 401  YDDLGKVFIY  HGSANGINTK  PTQVLKGISP  YFGYSIAGNM  DLDRNSYPDV

 451  AVGSLSDSVT  IFRSRPVINI  QKTITVTPNR  IDLRQKTACG  APSGICLQVK

 501  SCFEYTANPA  GYNPSISIVG  TLEAEKERRK  SGLSSRVQFR  NQGSEPKYTQ

 551  ELTLKRQKQK  VCMEETLWLQ  DNIRDKLRPI  PITASVEIQE  PSSRRRVNSL

 601  PEVLPILNSD  EPKTAHIDVH  FLKEGCGDDN  VCNSNLKLEY  KFCTREGNQD

 651  KFSYLPIQKG  VPELVLKDQK  DIALEITVTN  SPSNPRNPTK  DGDDAHEAKL

 701  IATFPDTLTY  SAYRELRAFP  EKQLSCVANQ  NGSQADCELG  NPFKRNSNVT

 751  FYLVLSTTEV  TFDTPYLDIN  LKLETTSNQD  NLAPITAKAK  VVIELLLSVS

 801  GVAKPSQVYF  GGTVVGEQAM  KSEDEVGSLI  EYEFRVINLG  KPLTNLGTAT

 851  LNIQWPKEIS  NGKWLLYLVK  VESKGLEKVT  CEPQKEINSL  NLTESHNSRK

 901  KREITEKQID  DNRKFSLFAE  RKYQTLNCSV  NVNCVNIRCP  LRGLDSKASL

 951  ILRSRLWNST  FLEEYSKLNY  LDILMRAFID  VTAAAENIRL  PNAGTQVRVT

1001  VFPSKTVAQY  SGVPWWIILV  AILAGILMLA  LLVFILWKCG  FFKRNKKDHY

1051  DATYHKAEIH  AQPSDKERLT  SDA
```

VLA-6

Domain structure

Seven tandem repeats in extracellular domain. The α6 chain is cleaved to yield a heavy and light chain joined by a disulphide bridge. Domains V, VI and VII each contain a divalent metal ion binding site. A 4th metal ion binding site resides between domains III and IV (residues 230–238). Platelet adhesion via VLA-6 is supported by Mn^{2+}, Co^{2+} and Mg^{2+} but not by Ca^{2+}, Zn^{2+} or Cu^{2+} [1].

1–23	Signal sequence
42–79	Repeat I
113–145	Repeat II
185–217	Repeat III
256–292	Repeat IV
314–352	Repeat V
375–411	Repeat VI
430–470	Repeat VII
1012–1037	Transmembrane domain
1038–1073	Cytoplasmic domain
899–903	Dibasic residues which may form the cleavage site for heavy and light chains.

Ten potential N-linked glycosylation sites.

Database accession numbers

	PIR	SWISSPROT	EMBL/GENBANK	REFERENCE
Human			X53586	8
			X59512	7

Amino acid sequence (β1)

See entry for VLA-1.

Alternative forms

In mouse, the α6 chain exists in two alternatively spliced forms which differ in the cytoplasmic domain. α6A is expressed in differentiated cells, whereas α6B is expressed only in embryonic stem cells which are undifferentiated and pleuripotent [2]. Two cytoplasmic domain variants of the α6 chain also occur in humans, designated 6A and 6B. The 6B form is 18 amino acids longer [7].

References
[1] **Sonnenberg, A. et al. (1988) Nature 360, 487–489.**
[2] Cooper, H.M. et al. (1991) J. Cell Biol. 115, 843–850.
[3] Shaw, et al. (1990) J. Cell Biol. 110, 2167–2174.
[4] de Curtis, et al. (1991) J. Cell Biol. 113, 405–416.
[5] Sonnenberg, A. et al. (1990) J. Cell Biol. 110, 2145–2155.
[6] Isberg, R.R. and Leong, J.M. (1990) Cell 60, 861–871.
[7] Hogervorst, F. et al. (1991) Eur. J. Biochem. 199, 425–433.
[8] **Tamura, R.N. et al. (1990) J. Cell Biol. 111, 1593–1604.**

Index

A

A CAM, 112-14
Adhesion molecule on glia, 22
α subunit(s) of integrin, 9, 10
α1 subunit, in integrin α1β1, 165-9
α2 subunit, in integrin α2β1, 170-2
α3 subunit, in integrin α3β1, 173-5
α4 subunit, in integrin α4β1, 176-80
α5 subunit, in integrin α5β1, 181-3
α6 subunit
    in integrin α6β1, 184-6
    in integrin α6β4, 79-82
α7 subunit, in integrin α7β1, 83-5
αIIb subunit, in integrin αIIbβ3, 135-9
αL subunit, in integrin αLβ2, 93-7
αM subunit, in integrin αMβ2, 105-8
αV subunit, in integrin αVβ3, 158-64
αX subunit, in integrin αXβ2, 89-92
AMOG, 22
ARC-1, see E-cadherin
Axonin-1, 151

B

B-cadherin, 23-5
β3, galactoprotein, 173-5
β subunit(s) of integrin, 9, 10, 11
β1 integrins, 9
β1 subunit
    in integrin α1β1, 165-9
    in integrin α2β1, 170-2
    in integrin α3β1, 173-5
    in integrin α4β1, 176-80
    in integrin α5β1, 181-3
    in integrin α6β1, 184-6
    in integrin α7β1, 83-5
    in integrin αVβ1, 163
β2 integrins, 9
β2 subunit
    in integrin αLβ2, 93-7
    in integrin αMβ2, 105-8
    in integrin αXβ2, 89-92
β3 integrins, 9
β3 subunit
    in integrin αIIbβ3, 135-9
    in integrin αVβ3, 158-64
β4 subunit, in integrin α6β4, 79
β5 subunit, in integrin αVβ5, 160-1, 163
β6 subunit, in integrin αVβ6, 161-2, 163
β8 subunit, in integrin αVβ8, 162-3, 163
BL-CAM, 34-5
B lymphocyte adhesion molecule, 34-5

C

C3b/C4b receptor, 39-42
Cadherin family, 6-8, *see also specific members*
    expression, regulation, 7
    function, 6
    signal transduction, 7
    structure, 6-7
Calcium-dependent lectin domain in selectins, 17
Carcinoembryonic antigen, see CEA
C-CAM, see Cell CAM
CD2, 26-7
CD4, 28-30
CD8, 31-3
CD11a/CD18, 93-7
CD11b/CD18, 105-8
CD11c/CD18, 89-92
CD18, see CD11
CD22, 34-5
CD23, 36-8
CD29, see CD49
CD31, 129-31
CD35, 39-42
CD36, 43-5
CD41/CD61, 135-9
CD42a/CD42b, 140-4
CD44, 46-8
CD49a/CD29, 165-9
CD49b/CD29, 170-2
CD49c/CD29, 173-5
CD49d/CD29, 176-80
CD49e/CD29, 181-3
CD49f/CD29, 184-6
CD51/CD61, 158-64
CD54, 74-7
CD56, see NCAM
CD58, 98-9
CD61, see CD41; CD51
CD62, see P-selectin
CEA (carcinoembryonic antigen), 49-51
    structure, 14, 49, 50

Index

Cell-CAM 105, 52-3
Cell-CAM 120/180, see E-cadherin
Collagen receptor, 170-2
Complement receptor, see CR
Complement regulatory protein domain in selectins, 17, 18
Constant region domain in immunoglobulin superfamily, 13
Contactin, 54-6
CR1, 39-42
CR3, 105-8
CR4, 89-92
C region domain in immunoglobulin superfamily, 13
CRP domain in selectins, 17, 18
C-type lectin domain in selectins, 17
Cytoadhesins, 9

D

D2-CAM, see NCAM
8D9, 123-5

E

E-cadherin (ARC-1; cell CAM120/180; L-CAM; liver cell adhesion molecule; uvomorulin), 57-9
  function, 6, 57
  K-CAM and, gene structure comparisons, 23
ECMR I, 170-2
ECMR II, 173-5
ECMR III, 46-8
EGF-domain in selectins, 17, 18, 19
ELAM-1, see E-selectin
EndoCAM, 129-31
Endothelial leukocyte adhesion molecule 1, see E-selectin
Epidermal growth factor domain in selectins, 17, 18, 19
E-selectin (ELAM-1; LECAM-2; endothelial leukocyte adhesion molecule 1), 60-3
  ligands, 18, 60

F

F3, 64-6
F11, 54-6
Fas, see Fasciclin
Fasciclin I, 67-8

Fasciclin II, 69-71
Fasciclin III, 72-3
FcεRII, 36-8
Fibronectin receptor
  integrin $\alpha 4\beta 1$ (VLA-4) as, 176
  integrin $\alpha 5\beta 1$ (VLA-5) as, 181-3
  integrin $\alpha V\beta 1$ as, 163

G

G4, neuronal-glial cell adhesion molecule, 123-5, 151
Galactoprotein $\beta 3$, 173-5
Glanzmann's thrombasthenia, 11, 135
Glia, adhesion molecule of, 22
GMP-140, see P-selectin
GP90^{HERMES}, 46-8
gp90 $^{MEL-14}$, see L-selectin
GPIb–GPIX, platelet glycoprotein, 140-4
GPIc–IIa, 181-3
GPIIb–IIIa, platelet glycoprotein, 135-9
GPIIIb, 43-5

H

H36 ($\alpha 7$ subunit) in integrin $\alpha 7\beta 1$, 83-5
H-CAM, 46-8
Hec7, 129-31
HUTCH-1, 46-8

I

ICAM-1, 74-7
  LFA-1 and, interactions, 93
ICAM-2, 77-8
IcIIa, 184-6
Ig, see Immunoglobulin
Immunoglobulin E receptor, low affinity, 36-8
Immunoglobulin superfamily, 13-16, see also specific members
  function, 15
  ligands, 14
  structure, 13-14
INCAM-1, see VCAM-1
In(lu)-related p80, 46-8
Integrin(s), family of, 9-12, see also specific members
  $\beta 1$, see $\beta 1$ subunit
  $\beta 2$, see $\beta 2$ subunit
  $\beta 3$, see $\beta 3$ subunit

Index

clinical aspects, 11
function, regulation, 11
ligands, 10-11
signal transduction, 11
structure, 9-10
Integrin $\alpha 1\beta 1$, 165-9
Integrin $\alpha 2\beta 1$, 170-2
Integrin $\alpha 3\beta 1$, 173-5
Integrin $\alpha 4\beta 1$, 176-80
Integrin $\alpha 5\beta 1$ ($\alpha_F\beta 1$), 181-3
Integrin $\alpha 6\beta 1$, 184-6
Integrin $\alpha 6\beta 4$ ($\alpha_F\beta 4$), 79-82
Integrin $\alpha 7\beta 1$, 83-5
Integrin $\alpha IIb\beta 3$, 135-9
Integrin $\alpha L\beta 2$, 93-7
Integrin $\alpha M\beta 2$, 105-8
Integrin $\alpha V\beta 5$ (and the $\beta 5$ subunit), 160-1, 163
Integrin $\alpha V\beta 6$ (and the $\beta 6$ subunit), 161-2, 163
Integrin $\alpha V\beta 8$ (and the $\beta 8$ subunit), 162-3, 163
Integrin $\alpha V\beta 1$, 163
Integrin $\alpha V\beta 3$, 158-64
Integrin $\alpha X\beta 2$, 89-92
Intercellular adhesion molecules, see ICAM

K
K-CAM, 23-5

L
L1, 86-8
L3T4, 28-30
LAM-1, see L-selectin
Laminin receptor
  CD49c/CD29 (VLA-3) as, 173-5
  CD49f/CD29 (VLA-6) as, 184-6
L-CAM, see E-cadherin
LECAM-1, see L-selectin
LECAM-2, see E-selectin
LECAM-3, see P-selectin
LEC-CAM-1, see L-selectin
Lectin domain, C-type, in selectins, 17
Leu2, 31-3
Leu-5, 26-7
Leu-8, see L-selectin
Leu-19, see NCAM
LeuCAM, subfamily, 9
LeuCAMc, 89-92
Leukocyte, extravasation, 19
Leukocyte adhesion deficiency, 11
Leukocyte adhesion molecules, see E-selectin; L-selectin
Leukocyte adhesion receptor Mo1, 105-8
Leukocyte adhesion receptor p150,95, 89-92
Leukocyte function antigens, see LFA
Leu M5, 89-92
Lewis^x determinant, sialyl, 18
LFA-1, 93-7
LFA-2, 26-7
LFA-3, 98-9
Liver cell adhesion molecule, see E-cadherin
L-MAG, 109, 110, 111
LPAM-2, 176-80
L-selectin (gp90^{MEL}-14; LAM-1; LECAM-1; LEC-CAM-1; leukocyte adhesion molecule-1; mLHR; TQ1), 100-2
  ligands, 18, 100
Ly-24, 46-8
Lyt2/Lyt3, 31-3

M
Mac-1, 105-8
MAG, 109-11
M-cadherin, 103-4
MFGM, 43-5
Milk fat globule membrane, 43-5
mLHR, see L-selectin
Mo1, leukocyte adhesion receptor, 105-8
Muscle-cadherin, 103-4
Myelin-associated glycoprotein, 109-11

N
N-cadherin, 112-14
N-CAL-CAM, 112-14
NCAM (CD56; D2-CAM; Leu-19; neural cell adhesion molecule; NKH1), 115-19
  function, 15, 115
  structure, 15, 116-18
NCAM-120, 116-17, 118
NCAM-125, 118
NCAM-140, 117, 118

Index

NCAM-180, 115, 116, 117-18, 118
Neural cadherin, 112-14
Neural cell adhesion molecule, *see* NCAM
Neuroglian, 120-2
Neuronal-glial cell adhesion molecule, 123-5
Neutrophil adherence receptor, 105-8
Ng-CAM, 123-5, 151
NILE, 86-8
NKH1, *see* NCAM

O
OKM5 antigen, 43-5
OX-8, 31-3

P
p80, in(lu)-related, 46-8
p85, 46-8
p150,95 leukocyte adhesion receptor, 89-92
PADGEM, *see* P-selectin
PAS IV, 43-5
P-cadherin, 126-8
Pgp-1, 46-8
Phosphorylation
  β subunit of integrin, 11
Placental cadherin, 126-8
Platelet endothelial cell adhesion molecule, 129-31
Platelet glycoprotein GPIb–IX, 140-4
Platelet glycoprotein GPIIb–IIIa, 135-9
Platelet glycoprotein Ia–IIa, 170-2
Platelet glycoprotein IV, 43-5
P-selectin (CD62; GMP-140; LECAM-3; PADGEM), 132-4
  ligands, 18, 132
  structure, 17, 133-4

R
R-cadherin, 145-7
Retinal cadherin, 145-7

S
Selectin family, 17-20, *see also specific members*
  function, 18, 19
  ligands, 18-19
  structure, 17-19
Sialyl Lewis[x] determinant, 18

T
T4, 28-30
T8, 31-3
T11, 26-7
TAG-1, 151-3
T-cadherin, 148-50
T cells
  CD2 and, 26
  CD8 and, 31
  CD11a/CD18 (LFA-1) and, 93
  CD44 and, 46
  CD49d/CD29 and, 176
  CD49e/CD29 and, 181
  CD54 and, 74
  receptor, CD4 as, 28
  selectin family and, 18
Thrombasthenia, Glanzmann's, 11, 135
T lymphocytes, *see* T cells
Tp50, 26-7
TQ1, *see* L-selectin
Truncated cadherin, 148-50
TSP 180, 79-82
Tyrosine phosphorylation, 176

U
Uvomorulin, *see* E-cadherin

V
Variable region domain in immunoglobulin superfamily, 13
Vascular cell adhesion molecule-1, *see* VCAM-1
VCAM-1 (INCAM-110; vascular cell adhesion molecule-1), 154-7
  structure, 14, 154-6
Very late antigens, *see* VLA
Vitronectin receptor, 158-64
VLA (very late antigens) protein subfamily, 9
VLA-1, 165-9
VLA-2, 170-2
VLA-3, 173-5
VLA-4, 176-80
VLA-5, 181-3
VLA-6, 184-6

W
W3/25, 28-30